STOLEN
Harvest

STOLEN
Harvest

The HIJACKING *of the* GLOBAL FOOD SUPPLY

by VANDANA SHIVA

SOUTH END PRESS   CAMBRIDGE, MA

Library of Congress Cataloging-in-Publication Data

Shiva, Vandana.
Stolen harvest : the hijacking of the global food supply / by Vandana Shiva.
    p. cm.
Includes index.
    ISBN 0-89608-608-9 (cloth : alk. Paper) -- ISBN 0-89608-607-0 (pbk. : alk. paper)
    1. Food industry and trade. 2. Big business. 3. Food supply. I. Title.

HD9000.5 .S454 1999
338.4'7664--dc21
ISBN-13 978-0-89608-607-4   99-016814

South End Press, 7 Brookline Street #1, Cambridge, MA 02139-4146
www.southendpress.org

CONTENTS

INTRODUCTION

*O*ver the past two decades every issue I have been engaged in as an ecological activist and organic intellectual has revealed that what the industrial economy calls "growth" is really a form of theft from nature and people.

It is true that cutting down forests or converting natural forests into monocultures of pine and eucalyptus for industrial raw material generates revenues and growth. But this growth is based on robbing the forest of its biodiversity and its capacity to conserve soil and water. This growth is based on robbing forest communities of their sources of food, fodder, fuel, fiber, medicine, and security from floods and drought.

While most environmentalists can recognize that converting a natural forest into a monoculture is an impoverishment, many do not extend this insight to industrial agriculture. A corporate myth has been created, shared by most mainstream environmentalists and development organizations, that industrial agriculture is necessary to grow more food and reduce hunger. Many also assume that intensive, industrial agriculture saves resources and, therefore, saves species. But in agriculture as much as in forestry, the growth illusion hides theft from nature and the poor, masking the creation of scarcity as growth.

These thefts have only stepped up since the advent of the globalized economy. The completion of the Uruguay Round of the General Agreement on Tariffs and Trade (GATT) in 1994 and the establishment of the World Trade Organization (WTO) have institutionalized and legalized corporate growth based on harvests stolen from nature and people. The WTO's Trade Related Intellectual Property Rights

1

Agreement criminalizes seed-saving and seed-sharing. The Agreement on Agriculture legalizes the dumping of genetically engineered foods on countries and criminalizes actions to protect the biological and cultural diversity on which diverse food systems are based.

The anti-globalization movement that started in response to GATT has grown tremendously, and I have been honored to have been part of it. My friends in the Third World Network, including Chakravarty Raghavan, and the tremendous people in the International Forum on Globalization have been a community of creativity and courage that has dared to challenge globalization at a time when history is supposed to have ended. Globally, we have seen the citizens movements against genetic engineering and corporate control over agriculture move concerns about genetic engineering from the fringe to the center stage of trade and economics. Whether at the St. Louis meeting on biodevastation or the Swiss or Austrian referenda on genetic engineering or the launch of the campaign for a Five Year Freeze on genetically engineered commerce in the United Kingdom, I have worked with some of the most courageous and creative people of our times who have taken on giant corporations and changed their fortunes. Corporations that have made governments their puppets and that have created instruments and institutions like the WTO for their own protection are now being held accountable to ordinary people.

A BRIEF HISTORY OF THE FIGHT TO SAVE THE STOLEN HARVEST

*I*n 1987, the Dag Hammarskjöld Foundation organized a meeting on biotechnology called "Laws of Life." This watershed event identified the emerging issues of genetic engineering and patenting. The meeting made it clear that the giant chemical companies were repositioning themselves as "life sciences" companies, whose goal was to control agriculture through patents, genetic engineering, and mergers. At that meeting, I decided I would dedicate the next decade of my life to find-

ing ways to prevent monopolies on life and living resources, both
through resistance and by building creative alternatives.

The first step I took was to start Navdanya, a movement for saving
seed, to protect biodiversity, and to keep seed and agriculture free of
monopoly control. The Navdanya family has started 16 community
seed banks in six states in India. Navdanya today has thousands of
members who conserve biodiversity, practice chemical-free agriculture, and have taken a pledge to continue to save and share the seeds
and biodiversity they have received as gifts from nature and their an-
cestors. Navdanya's commitment to saving seed means we cannot cooperate with patent laws, which make seed-saving a crime.

Seed patent laws, forced upon countries by WTO rules, are not the
only way in which the resources of the Third World poor are being stolen to generate profits for giant corporations. In 1994, the coastal communities of India invited me to support their struggle against industrial
shrimp farming, which was spreading like a cancer along India's
7,000-kilometer coastline. The Jaganathans, an amazing Gandhian
couple, had been leading a "shrimp *satyagraha*," or non-violent direct
action, to stop the devastation of coastal ecosystems and coastal communities. We joined forces with others like Bankey Behari Das of
Orissa, Tom Kochery of Kerala, Jesurithinam of Tamil Nadu, Claude
Alvares of Goa, and Jacob Dharmaraj in Andhra Pradesh to challenge
the shrimp-farming industry in a case that was heard before the Supreme Court of India in 1996. While the court ruled in our favor, commercial interests continue to attempt to subvert its judgement.

In August 1998, I witnessed the destruction of India's edible-oil
economy by the imposition of soybean oil, a pattern being replayed in
every sector of agriculture and the food economy. The women's
movement and farmers' movements resisted the imports of subsidized soybean oil to ensure that their livelihoods and their traditional
food cultures were not destroyed. In so doing, they demonstrated that
food free from genetic engineering is not a luxury for rich consumers.
It is a basic element of the right to safe, accessible, and culturally appropriate food.

On August 9, 1998, which is celebrated as Quit India Day in com-
memoration of the "Quit India" message given by Mohandas K. Gan-

dhi to the British, we started the "Monsanto, Quit India" campaign against the corporate hijacking of our seed and food. This movement against genetically engineered crops and food is now a global citizen's movement, involving farmers and consumers, activists and scientists. This book tells the stories of global corporations' destruction of food and agriculture systems as well as resistance to the destruction by people's movements.

These are exciting times. As the examples in this book show, it is not inevitable that corporations will control our lives and rule the world. We have a real possibility to shape our own futures. We have an ecological and social duty to ensure that the food that nourishes us is not a stolen harvest.

In this duty, we have the opportunity to work for the freedom and liberation of all species and all people. Something as simple and basic as food has become the site for these manifold and diverse liberations in which every one of us has an opportunity to participate—no matter who we are, no matter where we are.

The HIJACKING of the GLOBAL FOOD SUPPLY

*F*ood is our most basic need, the very stuff of life.

According to an ancient Indian Upanishad, "All that is born is born of *anna* [food]. Whatever exists on earth is born of *anna*, lives on *anna*, and in the end merges into *anna*. *Anna* indeed is the first born amongst all beings."[1]

More than 3.5 million people starved to death in the Bengal famine of 1943. Twenty million were directly affected. Food grains were appropriated forcefully from the peasants under a colonial system of rent collection. Export of food grains continued in spite of the fact that people were going hungry. As the Bengali writer Kali Charan Ghosh reports, 80,000 tons of food grain were exported from Bengal in 1943, just before the famine. At the time, India was being used as a supply base for the British military. "Huge exports were allowed to feed the people of other lands, while the shadow of famine was hourly lengthening on the Indian horizon."[2]

More than one-fifth of India's national output was appropriated for war supplies. The starving Bengal peasants gave up over two-thirds of the food they produced, leading their debt to double. This, coupled with speculation, hoarding, and profiteering by traders, led to skyrocketing prices. The poor of Bengal paid for the empire's war through hunger and starvation—and the "funeral march of the Bengal peasants, fishermen, and Artisans."[3]

Dispossessed peasants moved to Calcutta. Thousands of female destitutes were turned into prostitutes. Parents started to sell their children. "In the villages jackals and dogs engaged in a tug-of-war for the bodies of the half-dead."[4]

As the crisis began, thousands of women organized in Bengal in defense of their food rights. "Open more ration shops" and "Bring down the price of food" were the calls of women's groups throughout Bengal.[5]

After the famine, the peasants also started to organize around the central demand of keeping a two-thirds, or *tebhaga,* share of the crops. At its peak the Tebhaga movement, as it was called, covered 19 districts and involved 6 million people. Peasants refused to let their harvest be stolen by the landlords and the revenue collectors of the British Empire. Everywhere peasants declared, *"Jan debo tabu dhan debo ne "*—"We will give up our lives, but we will not give up our rice." In the village of Thumniya, the police arrested some peasants who resisted the theft of their harvest. They were charged with "stealing paddy."[6]

A half-century after the Bengal famine, a new and clever system has been put in place, which is once again making the theft of the harvest a right and the keeping of the harvest a crime. Hidden behind complex free-trade treaties are innovative ways to steal nature's harvest, the harvest of the seed, and the harvest of nutrition.

THE CORPORATE HIJACKING
OF FOOD AND AGRICULTURE

I focus on India to tell the story of how corporate control of food and globalization of agriculture are robbing millions of their livelihoods and their right to food both because I am an Indian and because Indian agriculture is being especially targeted by global corporations. Since 75 percent of the Indian population derives its livelihood from agriculture, and every fourth farmer in the world is an Indian, the impact of globalization on Indian agriculture is of global significance.

However, this phenomenon of the stolen harvest is not unique to India. It is being experienced in every society, as small farms and small farmers are pushed to extinction, as monocultures replace biodiverse crops, as farming is transformed from the production of nourishing and diverse foods into the creation of markets for genetically engineered seeds, herbicides, and pesticides. As farmers are transformed from producers into consumers of corporate-patented agricultural products, as markets are destroyed locally and nationally but expanded globally, the myth of "free trade" and the global economy becomes a means for the rich to rob the poor of their right to food and even their right to life. For the vast majority of the world's people—70 percent—earn their livelihoods by producing food. The majority of these farmers are women. In contrast, in the industrialized countries, only 2 percent of the population are farmers.

FOOD SECURITY IS IN THE SEED

F or centuries Third World farmers have evolved crops and given us the diversity of plants that provide us nutrition. Indian farmers evolved 200,000 varieties of rice through their innovation and breeding. They bred rice varieties such as Basmati. They bred red rice and brown rice and black rice. They bred rice that grew 18 feet tall in the Gangetic floodwaters, and saline-resistant rice that could be grown in the coastal water. And this innovation by farmers has not stopped.

Farmers involved in our movement, Navdanya, dedicated to conserving native seed diversity, are still breeding new varieties.

The seed, for the farmer, is not merely the source of future plants and food; it is the storage place of culture and history. Seed is the first link in the food chain. Seed is the ultimate symbol of food security.

Free exchange of seed among farmers has been the basis of maintaining biodiversity as well as food security. This exchange is based on cooperation and reciprocity. A farmer who wants to exchange seed generally gives an equal quantity of seed from his field in return for the seed he gets.

Free exchange among farmers goes beyond mere exchange of seeds; it involves exchanges of ideas and knowledge, of culture and heritage. It is an accumulation of tradition, of knowledge of how to work the seed. Farmers learn about the plants they want to grow in the future by watching them grow in other farmers' fields.

Paddy, or rice, has religious significance in most parts of the country and is an essential component of most religious festivals. The *Akti* festival in Chattisgarh, where a diversity of *indica* rices are grown, reinforces the many principles of biodiversity conservation. In Southern India, rice grain is considered auspicious, or *akshanta*. It is mixed with *kumkum* and turmeric and given as a blessing. The priest is given rice, often along with coconut, as an indication of religious regard. Other agricultural varieties whose seeds, leaves, or flowers form an essential component of religious ceremonies include coconut, betel, arecanut, wheat, finger and little millets, horsegram, blackgram, chickpea, pigeon pea, sesame, sugarcane, jackfruit seed, cardamom, ginger, bananas, and gooseberry.

New seeds are first worshipped, and only then are they planted. New crops are worshipped before being consumed. Festivals held before sowing seeds as well as harvest festivals, celebrated in the fields, symbolize people's intimacy with nature.[7] For the farmer, the field is the mother; worshipping the field is a sign of gratitude toward the earth, which, as mother, feeds the millions of life forms that are her children.

But new intellectual-property-rights regimes, which are being universalized through the Trade Related Intellectual Property Rights

Agreement of the World Trade Organization (WTO), allow corporations to usurp the knowledge of the seed and monopolize it by claiming it as their private property. Over time, this results in corporate monopolies over the seed itself.

Corporations like RiceTec of the United States are claiming patents on Basmati rice. Soybean, which evolved in East Asia, has been patented by Calgene, which is now owned by Monsanto. Calgene also owns patents on mustard, a crop of Indian origin. Centuries of collective innovation by farmers and peasants are being hijacked as corporations claim intellectual-property rights on these and other seeds and plants.[8]

"FREE TRADE" OR "FORCED TRADE"

*T*oday, ten corporations control 32 percent of the commercial-seed market, valued at $23 billion, and 100 percent of the market for genetically engineered, or transgenic, seeds.[9] These corporations also control the global agrochemical and pesticide market. Just five corporations control the global trade in grain. In late 1998, Cargill, the largest of these five companies, bought Continental, the second largest, making it the single biggest factor in the grain trade. Monoliths such as Cargill and Monsanto were both actively involved in shaping international trade agreements, in particular the Uruguay Round of the General Agreement on Trade and Tarriffs, which led to the establishment of the WTO.

This monopolistic control over agricultural production, along with structural adjustment policies that brutally favor exports, results in floods of exports of foods from the United States and Europe to the Third World. As a result of the North American Free Trade Agreement (NAFTA), the proportion of Mexico's food supply that is imported has increased from 20 percent in 1992 to 43 percent in 1996. After 18 months of NAFTA, 2.2. million Mexicans have lost their jobs, and 40 million have fallen into extreme poverty. One out of two peasants is not getting enough to eat. As Victor Suares has stated, "Eating more cheaply on imports is not eating at all for the poor in Mexico."[10]

In the Philippines, sugar imports have destroyed the economy. In Kerala, India, the prosperous rubber plantations were rendered unviable due to rubber imports. The local $350 million rubber economy was wiped out, with a multiplier effect of $3.5 billion on the economy of Kerala. In Kenya, maize imports brought prices crashing for local farmers who could not even recover their costs of production.

Trade liberalization of agriculture was introduced in India in 1991 as part of a World Bank/International Monetary Fund (IMF) structural adjustment package. While the hectares of land under cotton cultivation had been decreasing in the 1970s and 1980s, in the first six years of World Bank/IMF-mandated reforms, the land under cotton cultivation increased by 1.7 million hectares. Cotton started to displace food crops. Aggressive corporate advertising campaigns, including promotional films shown in villages on "video vans," were launched to sell new, hybrid seeds to farmers. Even gods, goddesses, and saints were not spared: in Punjab, Monsanto sells its products using the image of Guru Nanak, the founder of the Sikh religion. Corporate, hybrid seeds began to replace local farmers' varieties.

The new hybrid seeds, being vulnerable to pests, required more pesticides. Extremely poor farmers bought both seeds and chemicals on credit from the same company. When the crops failed due to heavy pest incidence or large-scale seed failure, many peasants committed suicide by consuming the same pesticides that had gotten them into debt in the first place. In the district of Warangal, nearly 400 cotton farmers committed suicide due to crop failure in 1997, and dozens more committed suicide in 1998.

Under this pressure to cultivate cash crops, many states in India have allowed private corporations to acquire hundreds of acres of land. The state of Maharashtra has exempted horticulture projects from its land-ceiling legislation. Madhya Pradesh is offering land to private industry on long-term leases, which, according to industry, should last for at least 40 years. In Andhra Pradesh and Tamil Nadu, private corporations are today allowed to acquire over 300 acres of land for raising shrimp for exports. A large percentage of agricultural production on these lands will go toward supplying the burgeoning food-processing industry, in which mainly transnational corporations are involved.

Meanwhile, the United States has taken India to the WTO dispute panel to contest its restrictions on food imports.

In certain instances, markets are captured by other means. In August 1998, the mustard-oil supply in Delhi was mysteriously adulterated. The adulteration was restricted to Delhi but not to any specific brand, indicating that it was not the work of a particular trader or business house. More than 50 people died. The government banned all local processing of oil and announced free imports of soybean oil. Millions of people extracting oil on tiny, ecological, cold-press mills lost their livelihoods. Prices of indigenous oilseed collapsed to less than one-third their previous levels. In Sira, in the state of Karnataka, police officers shot farmers protesting the fall in prices of oilseeds.

Imported soybeans' takeover of the Indian market is a clear example of the imperialism on which globalization is built. One crop exported from a single country by one or two corporations replaced hundreds of foods and food producers, destroying biological and cultural diversity, and economic and political democracy. Small mills are now unable to serve small farmers and poor consumers with low-cost, healthy, and culturally appropriate edible oils. Farmers are robbed of their freedom to choose what they grow, and consumers are being robbed of their freedom to choose what they eat.

CREATING HUNGER WITH MONOCULTURES

*G*lobal chemical corporations, recently reshaped into "life sciences" corporations, declare that without them and their patented products, the world cannot be fed.

As Monsanto advertised in its $1.6 million European advertising campaign:

> Worrying about starving future generations won't feed them. Food biotechnology will. The world's population is growing rapidly, adding the equivalent of a China to the globe every ten years. To feed these billion more mouths, we can try extending our farming land or squeezing greater harvests out of existing cultivation. With

the planet set to double in numbers around 2030, this heavy dependency on land can only become heavier. Soil erosion and mineral depletion will exhaust the ground. Lands such as rainforests will be forced into cultivation. Fertilizer, insecticide, and herbicide use will increase globally. At Monsanto, we now believe food biotechnology is a better way forward.[11]

But food is necessary for all living species. That is why the *Taittreya Upanishad* calls on humans to feed all beings in their zone of influence.

Industrial agriculture has not produced more food. It has destroyed diverse sources of food, and it has stolen food from other species to bring larger quantities of specific commodities to the market, using huge quantities of fossil fuels and water and toxic chemicals in the process.

It is often said that the so-called miracle varieties of the Green Revolution in modern industrial agriculture prevented famine because they had higher yields. However, these higher yields disappear in the context of total yields of crops on farms. Green Revolution varieties produced more grain by diverting production away from straw. This "partitioning" was achieved through dwarfing the plants, which also enabled them to withstand high doses of chemical fertilizer.

However, less straw means less fodder for cattle and less organic matter for the soil to feed the millions of soil organisms that make and rejuvenate soil. The higher yields of wheat or maize were thus achieved by stealing food from farm animals and soil organisms. Since cattle and earthworms are our partners in food production, stealing food from them makes it impossible to maintain food production over time, and means that the partial yield increases were not sustainable.

The increase in yields of wheat and maize under industrial agriculture were also achieved at the cost of yields of other foods a small farm provides. Beans, legumes, fruits, and vegetables all disappeared both from farms and from the calculus of yields. More grain from two or three commodities arrived on national and international markets, but less food was eaten by farm families in the Third World.

The gain in "yields" of industrially produced crops is thus based on a theft of food from other species and the rural poor in the Third

World. That is why, as more grain is produced and traded globally, more people go hungry in the Third World. Global markets have more commodities for trading because food has been robbed from nature and the poor.

Productivity in traditional farming practices has always been high if it is remembered that very few external inputs are required. While the Green Revolution has been promoted as having increased productivity in the absolute sense, when resource use is taken into account, it has been found to be counterproductive and inefficient.

Perhaps one of the most fallacious myths propagated by Green Revolution advocates is the assertion that high-yielding varieties have reduced the acreage under cultivation, therefore preserving millions of hectares of biodiversity. But in India, instead of more land being released for conservation, industrial breeding actually increases pressure on the land, since each acre of a monoculture provides a single output, and the displaced outputs have to be grown on additional acres, or "shadow" acres.[12]

A study comparing traditional polycultures with industrial monocultures shows that a polyculture system can produce 100 units of food from 5 units of inputs, whereas an industrial system requires 300 units of input to produce the same 100 units. The 295 units of wasted inputs could have provided 5,900 units of additional food. Thus the industrial system leads to a decline of 5,900 units of food. This is a recipe for starving people, not for feeding them.[13]

Wasting resources creates hunger. By wasting resources through one-dimensional monocultures maintained with intensive external inputs, the new biotechnologies create food insecurity and starvation.

THE INSECURITY OF IMPORTS

*A*s cash crops such as cotton increase, staple-food production goes down, leading to rising prices of staples and declining consumption by the poor. The hungry starve as scarce land and water are diverted to provide luxuries for rich consumers in Northern countries.

Flowers, fruits, shrimp, and meat are among the export commodities being promoted in all Third World countries.

When trade liberalization policies were introduced in 1991 in India, the agriculture secretary stated that "food security is not food in the *go-downs* but dollars in the pocket." It is repeatedly argued that food security does not depend on food "self-sufficiency" (food grown locally for local consumption), but on food "self-reliance" (buying your food from international markets). According to the received ideology of free trade, the earnings from exports of farmed shrimp, flowers, and meat will finance imports of food. Hence any shortfall created by the diversion of productive capacity from growing food for domestic consumption to growing luxury items for consumption by rich Northern consumers would be more than made up.

However, it is neither efficient nor sustainable to grow shrimp, flowers, and meat for export in countries such as India. In the case of flower exports, India spent Rs. 1.4 billion as foreign exchange for promoting floriculture exports and earned a mere Rs. 320 million.[14] In other words, India can buy only one-fourth of the food it could have grown with export earnings from floriculture.[15] Our food security has therefore declined by 75 percent, and our foreign exchange drain increased by more than Rs. 1 billion.

In the case of meat exports, for every dollar earned, India is destroying 15 dollars' worth of ecological functions performed by farm animals for sustainable agriculture. Before the Green Revolution, the byproducts of India's culturally sophisticated and ecologically sound livestock economy, such as the hides of cattle, were exported, rather than the ecological capital, that is, the cattle themselves. Today, the domination of the export logic in agriculture is leading to the export of our ecological capital, which we have conserved over centuries. Giant slaughterhouses and factory farming are replacing India's traditional livestock economy. When cows are slaughtered and their meat is exported, with it are exported the renewable energy and fertilizer that cattle provide to the small farms of small peasants. These multiple functions of cattle in farming systems have been protected in India through the metaphor of the sacred cow. Government agencies cleverly disguise the slaughter of cows, which would outrage many Indians, by calling it "buffalo meat."

In the case of shrimp exports, for every acre of an industrial shrimp farm, 200 acres of productive ecosystems are destroyed. For every dollar earned as foreign exchange from exports, six to ten dollars' worth of destruction takes place in the local economy. The harvest of shrimp from aquaculture farms is a harvest stolen from fishing and farming communities in the coastal regions of the Third World. The profits from exports of shrimp to U.S., Japanese, and European markets show up in national and global economic growth figures. However, the destruction of local food consumption, ground-water resources, fisheries, agriculture, and livelihoods associated with traditional occupations in each of these sectors does not alter the global economic value of shrimp exports; such destruction is only experienced locally.

In India, intensive shrimp cultivation has turned fertile coastal tracts into graveyards, destroying both fisheries and agriculture. In Tamil Nadu and Andhra Pradesh, women from fishing and farming communities are resisting shrimp cultivation through *satyagraha*. Shrimp cultivation destroys 15 jobs for each job it creates. It destroys $5 of ecological and economic capital for every dollar earned through exports. Even these profits flow for only three to five years, after which the industry must move on to new sites. Intensive shrimp farming is a non-sustainable activity, described by United Nations agencies as a "rape and run" industry.

Since the World Bank is advising all countries to shift from "food first" to "export first" policies, these countries all compete with each other, and the prices of these luxury commodities collapse. Trade liberalization and economic reform also include devaluation of currencies. Thus exports earn less, and imports cost more. Since the Third World is being told to stop growing food and instead to buy food in international markets by exporting cash crops, the process of globalization leads to a situation in which agricultural societies of the South become increasingly dependent on food imports, but do not have the foreign exchange to pay for imported food. Indonesia and Russia provide examples of countries that have moved rapidly from food-sufficiency to hunger because of the creation of dependency on imports and the devaluation of their currencies.

STEALING NATURE'S HARVEST

*G*lobal corporations are not just stealing the harvest of farmers. They are stealing nature's harvest through genetic engineering and patents on life forms.

Genetically engineered crops manufactured by corporations pose serious ecological risks. Crops such as Monsanto's Roundup Ready soybeans, designed to be resistant to herbicides, lead to the destruction of biodiversity and increased use of agrochemicals. They can also create highly invasive "superweeds" by transferring the genes for herbicide resistance to weeds. Crops designed to be pesticide factories, genetically engineered to produce toxins and venom with genes from bacteria, scorpions, snakes, and wasps, can threaten non-pest species and can contribute to the emergence of resistance in pests and hence the creation of "superpests." In every application of genetic engineering, food is being stolen from other species for the maximization of corporate profits.

To secure patents on life forms and living resources, corporations must claim seeds and plants to be their "inventions" and hence their property. Thus corporations like Cargill and Monsanto see nature's web of life and cycles of renewal as "theft" of their property. During the debate about the entry of Cargill into India in 1992, the Cargill chief executive stated, "We bring Indian farmers smart technologies, which prevent bees from usurping the pollen."[16] During the United Nations Biosafety Negotiations, Monsanto circulated literature that claimed that "weeds steal the sunshine."[17] A worldview that defines pollination as "theft by bees" and claims that diverse plants "steal" sunshine is one aimed at stealing nature's harvest, by replacing open, pollinated varieties with hybrids and sterile seeds, and destroying biodiverse flora with herbicides such as Monsanto's Roundup.

This is a worldview based on scarcity. A worldview of abundance is the worldview of women in India who leave food for ants on their doorstep, even as they create the most beautiful art in *kolams, mandalas,* and *rangoli* with rice flour. Abundance is the worldview of peasant women who weave beautiful designs of paddy to hang up for birds when the birds do not find grain in the fields. This view of abundance recognizes that, in giving food to other beings and species, we maintain conditions for our own food security. It is the recognition in the *Isho Upanishad* that the universe is the creation of the Supreme Power meant for the benefits of (all) creation. Each individual life form must learn to enjoy its benefits by farming a part of the system in close relation with other species. Let not any one species encroach upon others' rights.[18] The *Isho Upanishad* also says,

> a selfish man over-utilizing the resources of nature to satisfy his own ever-increasing needs is nothing but a thief, because using resources beyond one's needs would result in the utilization of resources over which others have a right.[19]

In the ecological worldview, when we consume more than we need or exploit nature on principles of greed, we are engaging in theft. In the anti-life view of agribusiness corporations, nature renewing and maintaining herself is a thief. Such a worldview replaces abundance with scarcity, fertility with sterility. It makes theft from nature a market imperative, and hides it in the calculus of efficiency and productivity.

FOOD DEMOCRACY

*W*hat we are seeing is the emergence of food totalitarianism, in which a handful of corporations control the entire food chain and destroy alternatives so that people do not have access to diverse, safe foods produced ecologically. Local markets are being deliberately destroyed to establish monopolies over seed and food systems. The destruction of the edible-oil market in India and the many ways through which farmers are prevented from having their own seed supply are small instances of an overall trend in which trade rules, property rights, and new technologies are used to destroy people-friendly and environment-friendly alternatives and to impose anti-people, anti-nature food systems globally.

The notion of rights has been turned on its head under globalization and free trade. The right to produce for oneself or consume according to cultural priorities and safety concerns has been rendered illegal according to the new trade rules. The right of corporations to force-feed citizens of the world with culturally inappropriate and hazardous foods has been made absolute. The right to food, the right to safety, the right to culture are all being treated as trade barriers that need to be dismantled.

This food totalitarianism can only be stopped through major citizen mobilization for democratization of the food system. This mobilization is starting to gain momentum in Europe, Japan, India, Brazil, and other parts of the world.

We have to reclaim our right to save seed and to biodiversity. We have to reclaim our right to nutrition and food safety. We have to reclaim our right to protect the earth and her diverse species. We have to stop this corporate theft from the poor and from nature. Food democracy is the new agenda for democracy and human rights. It is the new agenda for ecological sustainability and social justice.

1   *Taittreya Upanishad,* Gorakhpur: Gita Press, p. 124.
2   Kali Charan Ghosh, *Famines in Bengal, 1770–1943,* Calcutta: Indian Associated Publishing Company, 1944.
3   Bondhayan Chattopadhyay, "Notes Towards an Understanding of the Bengal Famine of 1943," *Transaction,* June 1981.
4   MARS (Mahila Atma Raksha Samiti, or Women's Self Defense League), Political Report prepared for Second Annual Conference, New Delhi: Research Foundation for Science, Technology, and Ecology (RFSTE), 1944.
5   Peter Custers, *Women in the Tebhaga Uprising,* Calcutta: Naya prokash, 1987, p. 52.
6   Peter Custers, p. 78.
7   Festivals like *Uganda, Ramanavami, Akshay Trateeya, Ekadashi Aluyana Amavase, Naga Panchami, Noolu Hunime, Ganesh Chaturthi, Rishi Panchami, Navartri, Deepavali, Rathasaptami, Tulsi Vivaha Campasrusti,* and *Bhoomi Puja* all include religious ceremonies around the seed.
8   Vandana Shiva, Vanaja Ramprasad, Pandurang Hegde, Omkar Krishnan, and Radha Holla-Bhar, "The Seed Keepers," New Delhi: Navdanya, 1995.
9   These companies are DuPont/Pioneer (U.S.), Monsanto (U.S.), Novartis (Switzerland), Groupe Limagrain (France), Advanta (U.K. and Netherlands), Guipo Pulsar/Scmins/ELM (Mexico), Sakata (Japan), KWS HG (Germany), and Taki (Japan).
10 Victor Suares, Paper presented at International Conference on Globalization, Food Security, and Sustainable Agriculture, July 30–31, 1996.
11 "Monsanto: Peddling 'Life Sciences' or 'Death Sciences'?" New Delhi: RFSTE, 1998.
12 ASSINSEL (International Association of Plant Breeders), "Feeding the 8 Billion and Preserving the Planet," Nyon, Switzerland: ASSINSEL.
13 Francesca Bray, "Agriculture for Developing Nations," *Scientific American,* July 1994, pp. 33–35.
14 *Business India,* March 1998.
15 T.N. Prakash and Tejaswini, "Floriculture and Food Security Issues: The Case of Rose Cultivation in Bangalore," in *Globalization and Food Security: Proceedings of Conference on Globalization and Agriculture,* ed. Vandana Shiva, New Delhi, August 1996.
16 Interview with John Hamilton, *Sunday Observer,* May 9, 1993.
17 Hendrik Verfaillie, speech delivered at the Forum on Nature and Human Society, National Academy of Sciences, Washington, DC, October 30, 1997.

18 Vandana Shiva, "Globalization, Gandhi, and Swadeshi: What is Economic Freedom? Whose Economic Freedom?" New Delhi: RFSTE, 1998.
19 Vandana Shiva, "Globalization, Gandhi, and Swadeshi."

SOY IMPERIALISM
and the DESTRUCTION
of LOCAL FOOD CULTURES

*T*he diversity of soils, climates, and plants has contributed to a diversity of food cultures across the world. The maize-based food systems of Central America, the rice-based Asian systems, the teff-based Ethiopian diet, and the millet-based foods of Africa are not just a part of agriculture; they are central to cultural diversity. Food security is not just having access to adequate food. It is also having access to culturally appropriate food. Vegetarians can starve if asked to live on meat diets. I have watched Asians feel totally deprived on bread, potato, and meat diets in Europe.

India is a country rich in biological diversity and cultural diversity of food systems. In the high Himalayan mountains, people eat pseudo-cereals such as amaranth, buckwheat, and chenopods. The people of the arid areas of Western India and semiarid tracts of the Deccan live on millets. Eastern India is home to rice and fish cultures, as are the states of Goa and Kerala. Each region also has its culturally specific edible oil used as a cooking medium. In the North and East it is mustard, in the West it is groundnut, in the Deccan it is sesame, and in Kerala it is coconut.

The diversity of oilseeds has also contributed to diversity of cropping systems. In the fields, oilseeds have always been mixed with cereals. Wheat is intercropped with mustard and sesame is intercropped with millets. A typical home garden could have up to 100 different species growing in cooperation.

The story of how the soybean displaced mustard in India within a few months of open imports is a story being repeated with different foods, crops, and cultures across the world, as subsidized exports from industrialized countries are dumped on agricultural societies, destroying livelihoods, biodiversity, and cultural diversity of food. The flooding of domestic markets with artificially cheap imports is stealing local markets and livelihoods from local farmers and local food processors. The expansion of global markets is taking place by extinguishing local economies and cultures.

"MUSTARD IS OUR LIFE"

*F*or Bengalis, Hilsa fish fried in mustard oil is the ultimate delight, and North Indians like their *pakoras* fried in it because of the unique taste and aroma. In the South, mustard seeds are the preferred seasoning for many dishes. Mustard oil is used as the cooking medium in the entire North Indian belt—the standard oil of Bihar, Bengal, Orissa and East Uttar Pradesh, used for flavoring and cooking.

Mustard, which was developed as a crop in India, is not just useful as an edible oil. It is an important medicine in the indigenous system of health care. It is used for therapeutic massages and for muscular and joint problems. Mustard oil with garlic and turmeric is used for rheumatism and joint pains. Mustard oil is also used as a mosquito repellent, a significant contribution in a region where the resurgence of malaria is responsible for the death of thousands.

There are many other personal and health care uses for mustard seeds and oil, and diverse varieties and species of mustard are grown and used for different purposes.[1] During the *Deepavali* celebration, mustard oil is used to light *diya* lamps. This is not just a celebratory tradition, but an ecological method of pest control at a time when the change in seasons causes an outbreak of disease and pests. The smoke from the mustard oil used to light the *deepavali* lamp acts as an environmental purifier and pest-control agent, reducing the spread of diseases that destroy stored grains and cleaning the atmosphere of homes and villages. As these mustard-oil lamps have been replaced by candles made of paraffin wax, an environmentally cleansing festival is transformed into an environmentally polluting one.

Indigenous oilseeds, being high in oil content, are easy to process at small-scale, decentralized levels with eco-friendly and health-friendly technologies. These oils are thus available to the poor at low cost. Hundreds and thousands of artisans are self-employed in rural India by extracting oil from locally produced crops for oil edible by humans and oil cake edible by cattle. The bulk of oilseed processing is done by over 1 million *ghanis* (expellers) and 20,000 small and tiny crushers that account for 68 percent of edible oils processed.[2] The oil extracted through these cold-pressing indigenous technologies is fresh, nutritious, unadulterated, and contains natural flavor.[3]

Women in the *bastis,* or slums, usually buy small quantities of mustard oil extracted on their local *ghani* in front of their eyes. This direct, community supervision over processing is the best guarantee for food safety. Yet these community-based systems of food and health safety were quickly dismantled in the name of food safety in 1998, when local processing of mustard oil was banned and free imports of soybean oil were installed in response to a mysterious contamination of Delhi's edible-oil supply.

The sudden lack of availability of mustard oil posed serious problems for poor women. Their children would not eat food cooked in imported palm oil or soybean oil, and were going to bed hungry. Being poor, they could not afford to buy the packaged oil that was the only form in which oil was available after the ban on local processors. For although the Chinese and Japanese eat soybean products as fermented foods, in most cultures outside East Asia, soybean products are not eaten. In spite of decades of promotion through free distribution in schools, soybean has not been adopted in India as a preferred choice for either oil or protein.

THE DROPSY EPIDEMIC

*D*uring August 1998, a tragedy unfolded in Delhi due to a massive adulteration of mustard oil with seeds of the weed *Argemone mexicana,* as well as other adulterants such as diesel, waste oil, and industrial oil.

Consumption of the adulterated oil had led to an epidemic of what was called "dropsy" and referred to a range of signs and symptoms affecting multiple organs and systems. These included nausea, vomiting, diarrhea, abdominal swelling, liver toxicity, kidney damage, cardiotoxicity, breathlessness due to retention of fluids in the lungs, and death due to heart failure. The link between dropsy and adulterated edible oil was first established by an Indian doctor in Bengal in 1926. By early September 1998, the official death toll was 41, and 2,300 people had been affected.

Mustard-oil sales were banned in Delhi, Assam, Bihar, Haryana, Madhya Pradesh, Orissa, Uttar Pradesh, West Bengal, Arunachal Pradesh, Sikkum, Tripura, and Karnataka. In July, India announced that it would import 1 million tons of soybeans for use as oilseeds, over the protests of citizen groups and the Agriculture Ministry, which challenged the necessity and safety of the imports. Later, free imports of soybeans were instituted. Not only was there no guarantee that these soybeans would not be contaminated with genetically engineered soybeans, the moves profoundly jeopardized the local oil-processing industry and with it the food culture and economy that depended on it.

On September 4, the government banned the sale of all unpackaged edible oils, thus ensuring that all household and community-level processing of edible oils stopped, and edible oil became fully industrialized. The food economy of the poor, who depend on unpackaged oil since it is cheaper and they can buy it in small quantities, was completely destroyed.

The adulteration that triggered these dire effects remains mysterious in origin. First, in the past local traders had adulterated particular brands of oils in remote and marginalized regions to cheat consumers in a way that would go unnoticed; however, the mustard-oil adulteration affected nearly all brands, and India's capital, Delhi, was the worst-affected region. Such an adulteration triggered an immediate response and could not have been initiated by an individual local trader.

Second, while corrupt traders had adulterated mustard oil with argemone in the past, before the 1998 tragedy, the adulterating agent was never found to be more than 1 percent of the oil. This time, contaminated oil contained up to 30 percent argemone and other agents. The high level of adulteration with argemone and other toxic substances such as diesel and waste oil clearly indicated that the tragedy was not the result of the normal business of adulteration.

According to the health minister of Delhi, the adulteration was not possible without an organized conspiracy. It was done in such a way that it could kill people quickly and conspicuously, and an immediate ban on mustard oil and free import of soybeans and other oilseeds for oil became inevitable. The Rajasthan Oil Industries Association claimed that a "conspiracy" was being hatched to undermine the mustard-oil trade, and felt that "invisible hands of the multinationals" were involved.

MULTINATIONAL COMPANIES GAIN FROM THE MUSTARD-OIL TRAGEDY

*D*uring the oil crisis, the Indian soybean lobby organized a major conference, "Globoil India 98," to promote the globalization and monoculturization of India's edible-oil economy. The U.S. Soybean Association was present at this conference to push for soybean imports.[4] According to *Business Line*, "U.S. farmers need big new export markets.... India is a perfect match."[5]

Multinational companies (MNCs) did gain from the mustard-oil tragedy. The ban on local processing has destroyed the domestic, small-scale edible-oil economy. It has criminalized the small-scale oil processor. It has criminalized the small trader. And it has destroyed the local market for farmers. Mustard prices have crashed from Rs. 2,200 to Rs. 600-800 per 100 kilograms.

The dangers of this destruction are tremendous. If traders cannot sell mustard oil, they will not buy mustard from farmers, and farmers will stop growing mustard. This will lead to the extinction of a crop that is the very symbol of Spring. Once mustard oil has gone out of cultivation, even after the ban is lifted on mustard oil, we will be forced to continue an enforced dependence on soybeans for edible oil.

Calgene, now owned by Monsanto, has patented the Indian mustard plant, the *India brassica.* If India wanted to reintroduce mustard later, it would have to depend on genetically engineered, patented mustard varieties. Farmers and consumers would be dependent on Monsanto for patented seeds of both soybean and mustard.

Such a reliance on imported oilseeds can easily trigger violence and instability. The food riots in Indonesia in the late 1990s were largely based on the fact that Indonesia had been made cripplingly dependent on imported soybeans for oil. When the Indonesian currency collapsed, the price of cooking oils shot up, and violence was the result.

Nor does the destruction of the domestic oil industry ensure greater food safety, as is argued by the government. It is an established fact that U.S. exports are heavily adulterated through what has been called purposeful contamination, or "blending." The toxic weed parthenium, which has spread across India, has been traced to wheat shipments from the United States.

More significantly, the adulteration of genetic engineering takes place at the genetic level and is hence invisible. Instead of toxic seeds like those of argemone being added *externally,* genetic engineering in effect allows food adulteration to be done *internally* by introducing genes for toxins from bacteria, viruses, and animals into crops. Genetic engineering is adulterating foods with toxins from rats and scorpions.

It is estimated that over 18 million acres were planted with genetically engineered Roundup Ready soybeans in 1998. The soybeans are engineered by Monsanto to contain a bacterial gene that confers tolerance to the herbicide Roundup, also manufactured by Monsanto. This soybean has been genetically engineered not in order to improve its yield or healthfulness. The sole purpose of Roundup Ready soybeans is to sell more chemicals for seeds tailored to these chemicals.

The United States has been unable to sell its genetically engineered soybeans to Europe because of European consumers' demands that such foods be labeled, something that is ardently opposed by agribusiness interests and their allies. According to former U.S. president Jimmy Carter, such labeling would make U.S. exports rot at ports around the world. (A wide-ranging coalition of U.S. scientists, health professionals, consumers, farmers, and religious leaders have filed a lawsuit demanding mandatory labeling.)

U.S. companies are therefore desperate to dump their genetically engineered soybeans on countries such as India. The mustard-oil tragedy is a perfect "market opening." For while the Indian government lost no time imposing packaging and labeling restrictions on the indigenous edible-oil industry, it has taken no steps to require segregation and labeling of genetically engineered soybeans.

A new soybean-futures exchange has been opened in India. According to Harsh Maheshwari of the Soya Association, the most conservative estimate of its activity is a turnover of $2.3 billion. Some say it will be five times more. The Council for Scientific Research and the Technology Mission on oilseeds have announced steps to promote the use of soybeans for food. Every agency of government in the United States and India is being used by the soybean lobby to destroy agricultural and food diversity in order to spread the soybean monoculture.

While the profits for agribusiness grow, the prices U.S. farmers receive for soybeans have been crashing. Both U.S. farmers and Indian farmers are losers in a globalized free-trade system that benefits global corporations.

GLOBAL MERCHANTS OF SOYBEANS

*I*n 1921, 36 firms accounted for 85 percent of U.S. grain exports. By the end of the 1970s, six giant "Merchants of Grain" controlled more than 90 percent of exports from the United States, Canada, Europe, Argentina, and Australia. Today, Cargill and Continental each control 25 percent of the grain trade.

Referring to this concentration of power, former Representative James Weaver (D-OR) said,

> These companies are giants. They control not only the buying and the selling of grain but the shipment of it, the storage of it, and everything else. It's obscene. I have rallied against them again and again. I think food is the most—hell, whoever controls the food supply has really got the people by the scrotum. And yet we allow six corporations to do this in secret. It's mind-boggling![6]

The United States is the world's biggest producer of soybeans, an East Asian crop that is also the United States' biggest export commodity. Twenty-six percent of U.S. acreage is under soybean cultivation. This production doubled between 1972 and 1997, from 34.6 million to 74.2 million metric tons. More than half of this crop is exported as soybeans or as soybean oil.

The U.S. acreage planted with genetically engineered soybeans has shot up from 0.5 million hectares in 1996 to 18 million hectares in 1998, accounting for 40 percent of the country's genetically engineered crops.[7] It is thus becoming inevitable that conventional soybeans will be mixed with genetically engineered soybeans in export shipments.

In the United States, soybeans are used for cattle feed, fish feed, adhesives, pesticides, plastics, solvents, soaps, paints, and inks.[8] Eighty percent of industrially processed foods now have soybeans in them, as European consumers discovered when they tried to boycott foods with Monsanto Roundup Ready soybeans.

Brazil follows the United States in soybean production, producing 30.7 million metric tons in 1997. Argentina is the third-biggest producer. Acreage in Argentina under soybean cultivation has increased from none in the 1960s to nearly 7 million hectares in 1998, with more than half planted with transgenic varieties. India's acreage under soybean cultivation has also increased from zero in the 1960s to nearly 6 million hectares in 1998.

The soybean trade, like trade in other agricultural commodities, is controlled by six Merchants of Grain: Cargill, Continental (now owned by Cargill), Louis Dreyfus, Bunge, Mitsui Cook, and Andre & Company.[10] These companies also control the storage and transport facilities, and hence the prices of commodities.

SOYBEAN PATENTS AND
SEED MONOPOLY

*N*ot only is the soybean trade controlled by multinational corporations; soybean cultivation is becoming increasingly monopolized through control over the seed itself.

Monsanto has bought up the seed business of corporations such as Cargill, Agracetus, Calgene, Asgrow Seed, Delta and Pine Land, Holden, Unilever, and Sementes Agrocetes. It owns the broad species patents on soybean. A subsidiary of W.R. Grace, Agracetus owns patent on all transgenic soybean varieties and seeds, regardless of the genes used, and all methods of transformation.

Agracetus's extraordinarily broad soybean patent has been challenged by Rural Advancement Foundation International, a public-interest group. Dr. Geoffrey Hawtin, director-general of the International Plant Genetic Resources Institute in Rome, Italy, expressed his concern at such patenting:

> The granting of patents covering all genetically engineered varieties of a species, irrespective of the genes concerned or how they were transferred, puts in the hands of a single inventor the possibility to control what we grow on our farms and in our gardens. At a stroke of a pen the research of countless farmers and scientists has potentially been negated in a single, legal act of economic hijack.[11]

While Monsanto had originally challenged the patent, it has withdrawn the challenge after buying Agracetus.

Monsanto also owns a patent on herbicide-resistant plants. This patent covers herbicide-resistant corn, wheat, rice, soybean, cotton, sugar beet, oilseed, rape, canola, flax, sunflower, potato, tobacco, alfalfa, poplar, pine, apple, and grape. It also covers methods for weed control, planting of seeds, and application of glyphosate (a herbicide). Thus Monsanto controls the entire production process of these plants, from breeding to cultivation to sale.

The Roundup Ready soybean has been genetically engineered to be resistant to Monsanto's broad-spectrum herbicide Roundup. The three new genes genetically engineered into the soybean—from a bacterium, a cauliflower virus, and a petunia—don't do a thing for the taste or nutritional value of the bean. Instead, the unusual genetic combination—which would never be created by nature—makes the soybean resistant to a weed-killer. Normally soybeans are too delicate to spray once they start sprouting from the ground. But now, since two of its products—the bean and the weed-killer—are so closely linked, Monsanto gets to sell more of both.[12] Monsanto claims this will mean more soybean yields from each crop, but they cannot guarantee it.

INDUSTRIAL PROCESSING

*F*rom seed to distribution to processing, soybeans are associated with concentration of power. While the oil content of coconut is 75 percent, ground nut 55 percent, sesame 50 percent, castor 56 percent, and niger 40 percent, the oil content of soybeans is only 18 percent. However, textbooks state that "soybean yields abundant supply of oil" and "soybeans have oil content higher than other pulses."[13]

Being low in oil content, soybean oil is extracted at large solvent-extraction plants. (Solvent-extraction was first applied in the United States to extract grease from garbage, bones, and cracking and packing house waste.) Chlorinated solvents such as chloroethylene are used to extract the oil.

Food safety is necessarily sacrificed in large-scale industrial processing since:

- the processing allows mixing of non-edible oils with edible oils,
- the processing is based on the use of chemicals,
- processing creates saturated fats,
- the long-distance transport lends itself to risks of adulteration, adds "food miles" in the form of CO_2 pollution, and contributes to climate change, and

- consumers are denied the right to know what ingredients have been used and what processing has been used to produce industrial oils.

ARE SOY PRODUCTS HEALTHY?

Soybeans and soybean products are being pushed as global substitutes for diverse sources of foods in diverse cultures. They are being promoted as substitutes for the diverse oilseeds and pulses of India and for cereals and dairy products worldwide. The American Soybean Association is promoting "analogue" dals—soybean extrusions shaped into pellets that look like black gram, green gram, pigeon pea, lentil, and kidney bean. The diet they envision would be a monoculture of soybean; only its appearance would be diverse.

However, even though the promotion of soybean-based foods is justified on grounds of health and nutrition, studies show that this sudden shift to soybean-based diets can be harmful to health. Soybean foods, in both raw and processed form, contain a number of toxic substances at concentration levels that pose significant health risks to humans and animals.

Soybeans have trypsin inhibitors that inhibit pancreatic processes, cause an increase in pancreatic size and weight, and can even lead to cancer.[14] In the United States, pancreatic cancer is the fifth most common fatal cancer, and its incidence is rising. The highest concentrations of trypsin inhibitors are found in soybean flour, which is a soy-based product that is not consumed in traditional soybean-eating cultures, which specialize in the consumption of fermented soybean products.[15]

Soybeans also have lectins that interfere with the immune system and the microbial ecology of the gut. When injected into rats, lectins isolated from soybeans were found to be lethal. When administered orally, these lectins inhibited rat growth.[16] Soybeans also contain phytic acid, which interferes in the absorption of essential minerals such as calcium, magnesium, zinc, copper, and iron. Given that deficiencies in calcium and iron are major symptoms of malnutrition in women and children in countries such as India, compromising the body's absorption of these essential minerals can have serious consequences.[17]

The most significant health hazard posed by diets rich in soybeans is due to their high estrogen content, especially in genetically engineered soybeans. The devastating impact of estrogenic compounds was highlighted when women born to mothers who took synthetic estrogens were found to have three times more miscarriages than other women and a greater incidence of a rare form of malignant vaginal cancer. Men born to mothers who took these synthetic estrogens had higher infertility levels than other men.[18]

Since soybeans are being used widely in all food products, including baby food, high doses of estrogen are being consumed by children, women, and men. Infants fed with soy-based formula are daily ingesting a dose of estrogens equivalent to that of 8 to 12 contraceptive pills.[19] According to New Zealand ecologist Richard James, soybean products are "unsafe at any speed and in any form."[20] The globalization of soybean-based foods is a major experiment being carried out on present and future generations. It is an unnecessary experiment, since nature has given us a tremendous diversity of safe foods, and diverse cultures have selected and evolved nutritious foods from nature's diversity.

During the mustard oil crisis in 1998, women from the slums of Delhi, organized by a women's group called "Sabla Sangh," invited me to discuss with them the roots of the crisis. They said that "Mustard is our life.... We want our cheap and safe mustard oil back." Ultimately, a women's alliance for food rights was formed. We held protests and distributed pure organic mustard oil as part of the *Sarson Satyagraha,* a program of non-cooperation against laws and policies that were denying people safe, cheap, and culturally appropriate foods.

The National Alliance for Women's Food Rights has challenged the ban on small-scale processing and local sales of open oil in the Supreme Court of India. We are building direct producer-consumer alliances to defend the livelihood of farmers and the diverse cultural choices of consumers. We protest soybean imports and call for a ban on the import of genetically engineered soybean products. As the women from the slums of Delhi sing, *"Sarson Bachao, Soya Bhagao,"* or "Save the Mustard, Dump the Soya."

The highest-level political and economic conflicts between freedom and slavery, democracy and dictatorship, diversity and monoculture have thus entered into the simple acts of buying edible oils and cooking our food. Will the future of India's edible-oil culture be based on mustard and other edible oilseeds, or will it become part of the globalized monoculture of soybean, with its associated but hidden food hazards?

1   Some of these diverse varieties include Indian mustard, *Brassica juncea*; black mustard, *Brassica nigra*; turnip rape; brown and yellow *Brassica campestris*; Indian rape; and rocket cross.
2   "Conspiracy in Mustard Oil Adulteration," *The Hindu,* September 17, 1998.
3   Status Paper on "Ghani Oil Industry," Mumbai: KVIC.
4   "Oilseeds Sector Needs to be Liberalized: U.S. Soya Body," *Economic Times,* September 22, 1998.
5   *Business Line,* October 12, 1998.
6   A.V. Krebs, "The Corporate Reapers: The Book of Agribusiness," Washington, DC: Essential Books, 1992.
7   Clive James, "Global Status of Transgenic Crops in 1997," ISAAA Briefs, Cambridge, MA: MIT Press, 1996. Also, Greg D. Horstmeier, "Lessons from Year One: Experience Changes How Farmers Will Grow Roundup Ready Beans in 98," *Farm Journal,* January 1998, p.16.
8   American Soybean Association, "Soy Stats, 1998."
10 A.V. Krebs.
11 Brian Belcher and Geoffrey Hawtin, "A Patent on Life Ownership of Plant and Animal Research," Ottawa, Canada: International Development Research Centre, 1991.
12 Vandana Shiva, "Mustard or Soya? The Future of India's Edible Oil Culture," Navdanya, 1998.
13 Dr. Irfan Khan, *Genetic Improvement of Oilseed Crops,* New Delhi: Ukaaz Publications, 1996, p. 334.
14 M.G. Fitzpatrick, "Report on Soybeans and Related Products: An Investigation into Their Toxic Effects," New Zealand: Allan Aspell and Associates, Analytical Chemists and Scientific Consultants, March 31, 1994, p. 5.
15 B.A. Charpentier and D.E. Lemmel, "A Rapid Automated Procedure for the Determination of Trypsin Inhibitor Activity in Soy Products and Common Food Stuffs," *Journal of Agricultural and Food Chemistry,* Vol. 32, 1984, p. 908.
16 I.E. Liener and M.J. Pallansch, "Purification of a Toxic Substance from Defatted Soy Bean Flour," *Journal of Biological Chemistry,* Vol. 197, 1952, p. 29.
17 S.L. Fitzgerald et al., "Trace Element Intakes and Dietary Phytat/Zn and Caz Phytate/Zn Millimolar Ratios in Periurban Guatemalan Women During the Third Trimester of Pregnancy," *American Journal of Clinical Nutrition,* Vol. 57, 1993, p. 725. See also J.W. Erdman and E.J. Fordyce, "Soy Products and the Human Diet," *American Journal of Clinical Nutrition,* Vol. 49, 1989, p. 725.
18 F.A. Kinil, "Hormone Toxicity in the Newborn," *Monographs on Endocrinology,* Vol. 31, 1990. See also R.J. Apfel and S.M. Fisher, *To Do No*

Harm: DES and the Dilemmas of Modern Medicine, New Haven: Yale University Press, 1984.

19 A. Axelsol et al., "Soya—A Dietary Source of the Non-Steroidal Oestregen Equal in Man and Animals," *Journal of Endocrinology,* Vol. 102, 1984, p. 49. See also K.D.R. Setchell et al., "Non-Steroidal Estrogens of Dietary Origin: Possible Roles in Hormone-dependent Disease," *American Journal of Clinical Nutrition,* Vol. 40, 1984, p. 569.

20 Richard James, "The Toxicity of Soy Beans and Their Related Products," unpublished manuscript, 1994, p. 1.

3

The STOLEN HARVEST
under the SEA

Worldwide, fish provide 17 percent of the animal protein in the human diet. Over 200 million people depend on fishing for their livelihoods.

Fish diversity is concentrated in tropical waters. The Indian and West Pacific Oceans contain an estimated 1,500 species of fish and over 6,000 mollusk species, compared with only 280 fish and 500 mollusk species in the Eastern Atlantic. The waters in Brazil are home to 3,000 freshwater fish species, and Thailand is home to more than 1,000 freshwater fish species.

While over 75 percent of the fish consumed by people comes from the harvest of wild species in natural ecosystems, industrial fish farming, or aquaculture, is the fastest-growing sector of global fish production, with shrimp aquaculture dominating the growth in tropical countries. Globally, more than half the shrimp and salmon consumed in the world is farmed, rather than caught in the wild.

The global fish catch has increased more than fourfold over the past 40 years. This massive harvest has been made possible by an explosion in industrial fishing fleets. Industrial fleets use massive drift nets to capture fish, using up to 3.5 million kilometers of synthetic netting every year, enough to circle the globe 88 times. Up to 50 percent of the fish caught in these "walls of death" drift nets are members of 200 non-commercial species.

As a result of these non-sustainable activities, an estimated 70 percent of the world's marine fish stocks are overfished or fully exploited, according to the United Nations Food and Agriculture Organization (FAO). Declining catches have destroyed more than 100,000 livelihoods and threaten millions more. With the collapse of the Canadian cod fishery, for instance, 80,000 fishermen and -women lost their livelihoods.

TURTLES AND SHRIMP

*I*n India, the turtle is considered sacred. It is one of the ten avatars, or incarnations, of Vishnu, the lord of creation and the maintainer. The *Satapatha Brahmana* states, "The Lord of progeny, having assumed the form of a tortoise, created offspring. He made the whole creation, hence the name Kurma given to the tortoise."[1]

In the myth of the churning of the oceans, the god Vishnu appeared in the form of the turtle to recover things lost in the deluge of the earlier era. The churning could take place only when Vishnu as turtle swam to the bottom of the ocean to serve as a pivot on which Mount Mandara rested, becoming a churning stick. The myth shows the significance of the turtle's role in sustaining life, and is the reason villagers along India's coasts relate to turtles with respectful reverence. Traditional fishing communities use non-violent technologies to ensure that marine species like turtles are not killed or hurt.

People and turtles have coexisted along India's coasts for centuries. But mechanized trawlers, introduced in the Indian waters over the past few decades through development financing and in the name of "modernization," profoundly threaten turtles. Industrial shrimp trawlers are capable of scraping one square kilometer of the seabed in ten hours, and an estimated 150,000 turtles drown each year when they are caught in the nets of large trawlers.

The Orissa Coast—the world's largest rookery of the endangered Olive Ridley turtles—is now famous for being their biggest grave. In November 1998, 26 dead turtles washed up on Orissa beaches. The next month, 652 dead turtles washed ashore, and by January 1999 the number of dead turtles had shot up to 4,682. Most of these were directly related to mechanized trawlers. In 1998, turtles did not come to the Gahirmata Beach in Orissa for mass nesting for the second year in a row.[2]

India is the seventh-largest producer of fish in the world and the second-biggest source of inland fish. Its 7,000 kilometer–long coastline supports the livelihood of millions of fishing and farming families. Until the end of the 1950s, the marine fish harvest in South Asia increased at a rate of 5 percent annually, despite the lack of new harvesting technologies. During this period, between 5,000 and 6,000 tons of prawns from India were exported to Burma, Thailand, and Malaysia every year, accounting for 25 to 30 percent of the annual export value of the shrimp trade.

Bottom-trawling was introduced to South Asia in the 1960s. In pursuit of shrimp, which usually are found in shallow waters, bottom-trawlers continuously rake the seabed, causing murky and turbid waters, and destroying the habitats of young bottom-dwelling fish and bottom-dwelling spawners.[3] By the late 1970s and early 1980s, the rate of growth of the marine fish harvest had dropped to 2 percent per year. However, despite the stagnation of the overall fishing economy, the exports of prawns—all destined for the Japanese and U.S. markets in frozen form—increased dramatically.

Trawler fleets use nets to scoop up whole shoals of fish, many of which are not of commercial value, although they are highly valuable to the ecosystem. Those species that do not have commercial value on global markets or are of the wrong size for standardized marketing and packaging are killed and thrown back into the sea. These fish are called "by-catch" and "discards." As *The Ecologist* reports, annual global discards in commercial fisheries have been conservatively estimated at 27 million tons, equivalent to over one-third the weight of all reported marine landings in commercial fisheries worldwide.[4] A study from Alaska suggests that Bering Sea red king crab discards amounted to more than five times the number of crabs actually landed. In the Norwegian cod fishery, the waste over one season in 1986–87 was 100,000 tons. In 1986–87, 2 billion kilograms of fin fish were dumped overboard.

Worldwide, the shrimp and prawn trawler fisheries are reported to have the highest level of discards of any fishery: about 16 million tons a year. In some shrimp fisheries, up to 15 tons of fish are dumped for every ton of shrimp landed. Most of this by-catch, turtles among it, is thrown back into the sea either dead or dying. These diverse species are the economic base for traditional fisherpeople and the ecological base that sustains the marine environment.

In terms of livelihoods, species diversity, and future sustainability, the technologies of industrial fisheries, which aim to maximize the commercial catch in the short run, are rather inefficient. Over-capitalized fisheries are collapsing in region after region. Nine of the world's major fishing grounds are threatened. Four have been "fished out" commercially. Total catches in the Northwest Atlantic have fallen by one-third over the past 20 years. In Newfoundland, fishing grounds have been closed indefinitely since 1992. In 1991, the FAO claimed that global fish catches would continue to increase, but even it now acknowledges that an estimated 70 percent of global fish stocks are "depleted" or "almost depleted" and that "the oceans' most valuable commercial species are fished to capacity."[5]

As marine ecology has degraded, the shrimp catch has also declined. In the major prawn-fishing area of southwest India, the catch dropped from 45,477 tons to 14,582 tons between 1973 and 1979. Trade sources also point to a shift in the composition of the export mix of prawns over time from the large species (*naran, kazhandan*) to the smaller varieties (*karikadi, poovalan*). These factors are widely accepted as indicators of overfishing.[6]

THE TURTLE VS. THE TRAWL

Since the 1970s, traditional fishing communities have been calling for a ban on mechanized trawlers in order to protect marine life and their livelihoods. They have called for Northern consumers, who are the beneficiaries of the export of Indian shrimp, to support this ban and boycott shrimp harvested by mechanized trawlers or farmed through non-sustainable aquaculture. This would, of course, involve a reduction in consumption by the rich and a reduction in global trade, but it would rejuvenate marine resources and the livelihoods of fishing communities.

Unfortunately, U.S. environmentalists' unawareness of the strong movements and stances of traditional fishing communities and environmental movements in India ultimately worsened the situation. While the U.S. environmental community took on the issue of turtle deaths due to shrimp trawling, it did not join Indian environmentalists in calling for a ban on trawling and consumer boycotts of shrimp. Instead, in the 1990s, U.S. environmental organizations called for the use of Turtle Exclusion Devices (TEDs) so that the turtles could escape if caught, advocating a ban on shrimp exports caught by vessels not using TEDs.

As stated in a brief prepared by U.S. environmental groups:

> The U.S. is one of the two largest consumers of shrimp products in the world, and its shrimp consumption is a major cause of turtle deaths. Given the causal connection between crimping and turtle mortality, the U.S. ability to reduce the impact of its shrimp consumption on sea turtles is critical to protecting endangered sea turtle populations. The use of TEDs in shrimp trawls that serve the large U.S. market represents the most environmentally sound and effective method available to the U.S. to protect these endangered species while allowing human crimping activity to continue relatively unimpeded.[7]

This shrimp ban was instituted by the United States in 1997. Asian countries, including India, Malaysia, Thailand, and Pakistan challenged the ban in a World Trade Organization (WTO) dispute. The resulting WTO ruling was indifferent to the environmental aspects of the ban, and merely focused on the trade dimensions. Since all environmental regulations restrict environmentally destructive commerce, they are trade-restrictive according to the WTO, hence illegal under the General Agreement on Tariffs and Trade.

Clearly, in this new era of defending the environment under globalization, a new solidarity and cooperation is needed between environmental movements in the South and in the North. Such a new solidarity would take into account that the real conflict over shrimp trawling is not between people and the turtle. Protecting the turtle should mean protecting traditional fishing communities and their culture of conservation, by strengthening environmental laws that protect both the environment and people. U.S. environmentalists' push for a limited ban on shrimp exports ultimately ended in the acceleration of environmental destruction. Since environmental deregulation is an essential part of trade liberalization, "free trade" and the protection of environment cannot coexist. If the turtle has to be saved, destructive trade and the use of destructive technologies need to end.

The WTO ruling is a victory for trading interests that have no loyalty to any country or any ecosystem. It is not a victory for India, because India is not the global shrimp industry: India is her coasts and marine line, her mountains and rivers, her farms and forests. India is the peasants and tribals and fishworkers whose resources and livelihoods are being destroyed by destruction of the environment. India is her turtles.

THE VIOLENCE OF THE "BLUE REVOLUTION"

*A*ccording to the International Food Policy Research Institute, "to meet the growing need for fish, the world will have to rely on aquaculture."[8]

The two primary justifications for industrial aquaculture are the crisis of depletion of marine resources and the crisis of malnutrition among the poor in the Third World. The World Bank and corporate investors, for instance, have promoted shrimp aquaculture as a way to meet the growing demand for shrimp in the face of declining catches from the wild.

Cultured-shrimp production has increased from 10 percent of total shrimp production in 1985 to 30 percent in 1992. Cultured shrimp contributed 12 million tons out of a total shrimp production of 98 million tons in 1989–91, and is expected to reach a production level of 15 to 20 million tons by 2010.[9] Though pushed by both national and international organizations as an answer to world food scarcity, particularly to the scarcity of proteins in the diets of the poor, in reality shrimp contributes little to the nutritional needs of the world's population, being a luxury item that is consumed mainly by the rich in the developed world.

Farming for prawn and fish is quite different from capturing prawn and fish that grow in the wild. The aquaculturist must maintain and run the prawn farm in the same way as an agricultural farm, paying attention to weather, nutrients, and feed to ensure a healthy crop. Sustainable aquaculture has been a part of sustainable agriculture in many ancient farming systems. However, modern industrial aquaculture, the "Blue Revolution," is of recent origin. As in the case of crop production, industrial fisheries and aquaculture consume more resources than they produce. According to Dr. John Kurien, in 1988 global shrimp aquaculture consumed 1.8 million tons of fish meal, derived from an equivalent of 900,000 tons (wet-weight) of fish. It is further estimated that by 2000, about 5.7 million tons of cultured fish will be produced in Asia. The feed requirements for this harvest will be on the order of 1.1 million tons of feed, derived from a staggering 5.5 million tons of wet-weight fish—nearly double the total marine fish harvested in India today.

Fish meal provides the crucial link between industrial aquaculture and industrial fisheries, since the fish used for fish meal are harvested from the sea through trawlers and purseiners, which are known to deplete marine stocks. This exposes the illogic of the World Bank argument that aquaculture moves away from hunting and gathering toward settled agriculture, and will reduce the pressure on marine resources.[10]

PUBLIC SUPPORT FOR
PRIVATE PROFITS

International aid to aquaculture increased from $368 million in 1978–84 to $910 million in 1988–93.[11] The World Bank has supported aquaculture since the 1970s, when it began providing loans to Asian and Latin American governments to develop shrimp ponds. The bank financed such development projects in Indonesia, the Philippines, Thailand, and Bangladesh. By the 1980s, the bank had broadened its support to include China, India, Brazil, Colombia, and Venezuela.[12] This investment emphasized infrastructure development, in the form of roads and refrigeration, paving the way for the expansion of industrial shrimp farming in the 1980s.[13]

In 1992 the bank invested $1.7 billion in agriculture and fisheries, of which India received $425 million for shrimp and fish culture. The bank noted that shrimp production in India, the world's largest producer and exporter of shrimp for the last two decades, was based on traditional shrimp-culture systems in which ponds were frequently used for paddy cultivation during the rainy season and converted to shrimp and fish culture for the rest of the year. According to the bank, as a result, shrimp yields were low (300 kilograms per hectare), reflecting poor infrastructure, low-density stocking, inadequate or no water exchange, a lack of feed, and low-level technology.[14] The bank argued that semi-intensive shrimp farming could help increase India's shrimp production, provide employment, and help the country earn much needed foreign revenue.[15]

In 1991 the Indian government set up the Marine Products Export Development Authority (MPEDA) to further support export-oriented aquaculture. MPEDA offered significant assistance and subsidies for aquaculture development in India.[16]

WESTERN LUXURY FOODS AND THIRD WORLD PRODUCERS

*W*hile Western countries such as the United States have highly productive and profitable shrimp farms, shrimp farming has not proliferated in the United States or in any other industrialized country. Instead, U.S. investment in aquaculture has grown in countries such as Mexico and Ecuador. In all, Western countries account for less than 25 percent of the world's shrimp production.[17]

This indicates that the environmental destruction caused by intensive shrimp farming is one of the major factors for its spread in Third World countries, even though the main consumers of shrimp live in affluent countries. In country after country where commercial shrimp farming has been tried, it has proved unsustainable. For this reason, this industry is known as a "rape and run" industry.

Taiwan was the world's largest producer of cultured shrimp until 1988, when a major disease outbreak led to a collapse from which Taiwan's shrimp industry has still not recovered. China then led world production until 1993, when its productivity dropped for similar reasons. Shrimp farms in India were subject to a major virus attack in 1994 and early 1995, which led the government to declare a "crop holiday" for the industry.

Presently, both production and market prices are controlled by disease outbreaks. But the shrimp market is unstable in other ways. The earnings of Third World producers are also dependent on the food fashions prevailing among the world's elite minority. When this minority moves on to other foods for either health or taste, the market will collapse.

DESTRUCTION OF THE MANGROVES: THE NURSERIES OF MARINE LIFE

*M*angroves play a crucial ecological role in coastal ecosystems by protecting against tropical rain storms, anchoring shifting mud and thus preventing erosion, and providing shelter and habitat for fish and other marine life.[18]

Shrimp ponds are the main cause of mangrove loss over the last few decades. Mangrove areas have dropped from 3,650 hectares in 1983 to 2,000 hectares in 1994 in Puttlam District, Sri Lanka.[19] In Vietnam, 102,000 hectares of mangroves were cleared for shrimp farming between 1983 and 1987.[20] Most of the 21,600 hectares of shrimp ponds in Ecuador were constructed in what were previously mangrove areas.[21] Of the 203,765 hectares of mangroves lost in Thailand between 1961 and 1993, 32 percent were converted into shrimp farms.[22]

The loss of mangroves leads to a depletion of marine resources, and hence declining catches for small fishing communities.

THE POLLUTION OF COASTAL WATERS

Shrimp farming requires four to six tons of feed per hectare. Only 17 percent of this feed is converted into shrimp biomass. The rest becomes waste, heavily contaminated with pesticides and antibiotics, which is flushed directly back into the sea or onto neighboring mangrove and agricultural lands. The shrimp pond is then refilled with new sea water. The high level of pollution resulting from this open drainage of effluents into both irrigation channels and the sea has resulted in fish mortality, the contamination of groundwaters, and various health hazards. [23]

There is also an increasing concern that cultured species may escape into the natural environment as well as into foreign environments, which may adversely affect the local aquatic ecology.[24]

SALINE DESERTS AND WATER FAMINE

Shrimp farming requires the pumping of sea water into ponds, since most of the shrimp species farmed require a salinity between 25 to 30 parts per trillion. A one-hectare industrial shrimp farm, for instance, requires 120,000 cubic meters of sea water every year. During the shrimps' growing period—between 120 and 150 days—salt water from the ponds seeps into neighboring agricultural farms and the water table.

The fact that fresh water from underground aquifers must be extracted for salinity control in the ponds intensifies the problem. Over the four-month growing period, roughly 6,600 cubic meters of fresh water are needed to dilute the sea water in a one-meter-deep, one-hectare pond. The aquifers left empty after these massive extractions are especially vulnerable to salt-water intrusion.

The salinization of the groundwater is creating a major drinking water crisis in coastal communities. At a 1997 public hearing held in Delhi, people from coastal villages reported how industrial shrimp farming had created water famines in areas formerly abundant with water.

Chandramohan of Jagidapattinam village in Ramnad district testified that

> Five to six years back, drinking water [and] growth of coconut and palm trees were not a problem. But since the establishment of 39 farms, drinking water has become a major problem. Trees have either withered or are cut to make way for the aqua farms. The villagers have to travel 10 kilometers to get water or have to pay five rupees per pot of water if it is transported by truck.

Govindamma of Kurru village in Nellore district reported,

> The village is surrounded by prawn farms on all four sides.... We have lost all our drinking water, where earlier there used to be nine wells in this area. We no longer live in this village as all the houses have collapsed because of dampness and salinity. Five hundred families have been displaced. Social tensions are created by the Aqua Companies, resulting in a fight between the Aqua Companies and the villagers, leading to three deaths in the village.

As coastal ecosystems are destroyed, and with them people's livelihoods, this additional burden is forcing families to migrate out of coastal villages.[25]

NO FOOD, NO WATER:
THE FEMINIZATION OF SUFFERING

*O*nce-fertile and -productive paddy fields are becoming what local people call "graveyards," unfit for agriculture. This is true not just for India but for other countries as well. In Bangladesh, home to intensive shrimp farms, the amount of rice production has dropped from 40,000 metric tons in 1976 to 36 metric tons in 1986. Thai farmers report similar losses due to the introduction of shrimp farms.

Women have been particularly affected by the proliferation of the shrimp industry. Land has become a scarce commodity. Fights take place between neighbors over patches of land on which to dry fish. In places where water is provided by tankers, competition for the water becomes yet another cause of social disruption, particularly between women.

In the village of Kurru in Nellore district, there was no drinking water available to the 600 fisherfolk, due to salinization of the drinking water. After local women held protests, the government started supplying drinking water in tankers. Each household gets two pots to drink, wash, and clean with. "Our men need ten buckets of water to bathe after their fishing trips. What can we do with two pots?" one woman asked. Women say they have to work four to six more hours daily to collect fuel and water as a result of the environmental destruction caused by shrimp farms.[26]

In another village in Andhra Pradesh, after two years of supplying drinking water to villagers in tankers, the state government decided to relocate 500 families. Still, there are a number of regions where people are left with no option but to use salt water for their crops and everyday needs.

The contaminated drinking water has led to numerous cattle deaths. There has also been a considerable decline in the growth of fodder. Two hundred head of cattle have died in Kurru village alone since the advent of commercial shrimp culture.

Where shrimp farms have been set up, the fish have left for deeper and calmer waters. According to fisherfolk, the amount of fish they used to catch in four hours before the advent of the industry today takes eight hours to catch.

If all the costs of shrimp farming are taken into account, it is clear that this farming is not sustainable. It poses a threat to coastal ecosystems and the survival of coastal communities. Because of this threat, in 1994 Indian environmentalists and coastal communities filed a public-interest suit in the Supreme Court of India, challenging industrial shrimp farming's destruction of coastal ecosystems and coastal peoples' livelihoods. In 1995 the court appointed an expert committee to look into the social and ecological costs of aquaculture.

SUSTAINABLE PRAWN CULTURE

*T*raditional systems of aquaculture, which have been used for over 500 years, though diverse, have some common features. They are based on local farming systems, have little adverse impact on the local ecology, and ensure the conservation and continuation of the various life forms present in the ecosystem. They are as profitable as the more intensive, industrial systems of commercial aquaculture. These traditional systems are responsible for India's status as the world's biggest producer of shrimp, and have provided domestic and local food security to the farmers and fisherfolk in the coastal regions.

The *bheri* system of aquaculture, for example, was developed in the tidal mudflats and swamp-marsh areas of the Upper and Lower Sunderbans in West Bengal. These irregular-shaped and -sized *bheris* range from 2 hectares to 267 hectares. There are two types: seasonal and perennial. The seasonal *bheris* are used from November to December, and then allowed to dry in the sun until the following season. In the perennial *bheris,* found exclusively in the high-salinity zones where no paddy is grown, fish and shrimp are raised throughout the year.

In Orissa, traditional aquaculture ponds called *gheris* are located near estuaries, seashores, and around lakes. They are constructed with bamboo sticks held in place by rope, while nets are used to capture and contain prawn and fish. The tides force fish, prawn, and other aquatic organisms into the nets. Once caught in the *gheri*, they are unable to escape, and are fed by food brought in by the tidal waters. Once the prawn and fish mature they are harvested. Modern *gheris* now provide some artificial feed to obtain quicker results.

Traditional shrimp farming and aquaculture have been practiced in Kerala's low-lying backwaters for centuries. In the seasonal fields, paddy is cultivated during the monsoon months (July-October), and prawn/fish cultivated during the rest of the year when the fields become inundated with saline waters. For rice cultivation, raised beds allow exposure to the sun and allow excess salt to seep out of the soil. Paddy seeds are sown and covered with coconut leaves. The field is completely filled with water once the roots of the rice seeds have stabilized. The backwaters help provide fertility to the soil through the nutrients and minerals that are washed in with the water. At harvest time, the upper portion is cut, and the rest is left behind for prawn and fish cultivation. The rice harvest is often consumed by the farmers themselves, with some rice sold on local markets.

For fish farming, sea water from the high tide is allowed onto the fields to stock the farms with juvenile shrimp and other fish. When the tide begins to recede, a closely knotted screen made of split bamboo is inserted across the gate, allowing the water out and trapping the juvenile shrimp in the field. This entrapment is continued at every high tide throughout the period of operation. Harvesting begins in mid-December. The final harvesting is done at the end of the season by sluice or cast net and by hand.

Paddy farmers often lease their land for prawn cultivation to more skilled prawn/fish farmers. However, now some paddy farmers are reluctant to do so since prawn farmers have started using artificial feed and chemicals, which affect the productivity of the paddy.

Generations of fisherfolk have been catching fish through the use of hand-constructed nets. Some traditional netting techniques can be carried out by a single person and can fetch anywhere between Rs. 100 and Rs. 200 per day. The fisherfolk usually follow traditional methods using astronomy and tidal readings to select the best time of the month (usually 15 days) to fish. Of the 15 days, five to six are considered to be particularly ideal for fishing. Fishing is carried out throughout the year in the sea, backwaters, canals, and ponds.

Other traditional systems of farming shrimp and fish include the *thappal,* which means "to search" in Malayalam. During high tide, fisherfolk use their hands to feel and search for prawns, oysters, and fish that may have been swept in toward the shore. The catch is placed in a bowl or pot filled with saline water. A technique associated with the *thappal* is the use of a mat made from dried grass and touch-me-nots, which are intertwined with rice grains placed on top of the mat. The mat with the grain is submerged in the water. The grain attracts prawns, which become trapped in the mat. These and other techniques of procuring prawn and fish have helped to sustain the livelihoods of coastal people for centuries.

THE SECOND "BLUE REVOLUTION"

*A*bout 50 labs around the world are conducting research on transgenic fish. Most of this research focuses on engineering rapid growth and cold-tolerance. A/F Protein, based in Canada and the United States, has engineered Atlantic salmon with a growth hormone gene that reportedly makes it grow to market size in 12 to 18 months instead of the usual three years. The company has patents on the gene and transformation method, and its genetically engineered salmon is called Biogrow.[27] In Scotland, Otter Ferry Salmon of Strathclyde is also experimenting with salmon engineered for faster growth. In Chile, a consortium of business interests wants to commercialize production of transgenic fish, which are supposed to grow ten times faster than normal.

While genetic engineering, like industrial aquaculture, is promoted to increase fish production, because of its ecological risks, it could in fact deplete fish stocks. For instance, the faster-growing transgenic fish may require more feed in order to grow at the increased rate. Transgenic fish with anti-freeze genes meant to tolerate colder sea water than their non-engineered relatives could displace other species.

The introduction of new genes could impact other physiological processes. For example, when fed a high-protein diet, transgenic pigs containing human or bovine growth hormone genes exhibited faster growth. However, females were sterile, and animals of both sexes were lethargic, exhibited muscle weakness, and had a propensity to develop arthritis and gastric ulcers.[28]

Transgenic fish could ruin aquatic ecosystems by preying on and outcompeting native species. Engineered fish could breed with wild fish and destroy diversity. Transgenic fish need to be considered as a special case of exotic fish. Introductions of exotics can have unpredictable and serious impact. Peter Moyle of the University of California at Davis has called the displacement of native species by the introduction of exotic species the "Frankenstein Effect."[29]

Examples of the Frankenstein Effect are the introduction of blue tilapia into Lake Effie in Florida and the introduction of opossum shrimp in Flathead Lake in Montana. When the tilapia was introduced in 1970, it consisted of less than 1 percent of the total weight (biomass) of fish in Lake Effie. By 1974, the blue tilapia accounted for more than 90 percent of the fish biomass.

Between 1968 and 1975, opossum shrimp were introduced into several lakes upstream from Flathead Lake to improve food sources for Kakonee salmon. However, the opposite happened. The shrimp were voracious predators of zooplankton, which is an important food source for the salmon. Zooplankton populations declined to 10 percent of their former levels, and the salmon catch plummeted. Before 1985, the annual salmon catch was 100,000. Only 600 were caught in 1987. There were no reported catches in 1989.

The release of genetically engineered fish, via the Second Blue Revolution, could prove equally disastrous socially and ecologically. Genetically engineered fish, offered as a new miracle in fisheries, intensifies the one-dimensional trajectory of the Blue Revolution to breed fish for higher production and faster growth. We can therefore expect that the devastation already experienced in the case of the Blue Revolution will be intensified and accelerated in the Second Blue Revolution.

THE LONG ROAD TO ENVIRONMENTAL JUSTICE

*I*n 1996, in response to a suit filed by Indian environmentalists and coastal communities, the Supreme Court of India ordered the removal of all shrimp aquaculture in the coastal regulation zones, comprising the coastal ecosystems of Bengal, Orissa, Andhra Pradesh, Tamil Nadu, Kerala, Karnataka, Goa, Maharashtra, and Gujurat.

The court ruled that "no aquaculture industry, whether it is intensive, semi-intensive, extensive, or semi-extensive, will be permitted. The only activity which will be permitted is traditional and improved traditional." By the end of March 1997, all aquaculture industries in the area were to be completely removed, and the aquaculture workers were to be paid retrenchment compensation plus six years of wages. The farmers of the area were to be compensated for their losses. The court ordered that the federal government designate an authority to carry out the far-reaching, landmark ruling. The court thus upheld the value of life above the value of dollars earned from shrimp exports.

According to one leading financial daily, undoing the judgement was a major priority for the government. Indeed, the government, along with business interests, has succeeded in preventing the ruling from coming into force. Shrimp farms continue to operate in contempt of court orders.

Environmentalists and coastal communities have organized a massive national and international mobilization to prevent a complete undoing of the historic Supreme Court judgement. However, the fundamental rights and freedoms of the poor coastal communities are under permanent threat because of the dollar power of the shrimp industry. It is these communities that are paying the real price for increased shrimp consumption—with their livelihoods and their freedom.

On the 1997 anniversary of India's independence day, August 15, while official India mouthed empty rhetoric and radicals staged a "Black Flag" demonstration against government failures, coastal villagers, under the leadership of the National Action Committee against

Coastal Industrial Aquaculture, marched to banned shrimp farms, proudly carrying the Indian tricolor flag and singing the national anthem. From the coast of India a new meaning is being given to freedom, both for the people and the country.

For the victims of the aquaculture industry, Independence Day was a day for celebrating and asserting their sovereignty over their natural resources and their livelihoods. It was a day for re-committing themselves to continuing their struggle to free the coast from the destructive aquaculture industry. It was a day for condemning the attempts by the government, politicians, and industrialists to subvert the Supreme Court judgement that has defended their rights and their coast.

This new struggle for a free India is appropriately beginning at India's social and environmental margins—from the coasts, led by women, traditional fishworkers, the landless, and small peasants. In the margins, a new India is being born—an India built on the principles of sustainability and justice, of peace and harmony, of democracy and diversity.

This second freedom struggle has just begun.

1   The first mention of this incarnation is found in the *Satapatha Brahmana*; it is also mentioned in the *Mahabaratha* (1.18), the *Ramayana* (1.45), and the *Puranas* (*Agni Purana,* ch. 3; *Kurma Purana,* ch. 259; *Vishnu Purana* 1.9; *Padma Purana* 6.259; as well as in the *Bhagavata Purana*).

2   Shri Banka Behari Das, "Serious Trouble for Bhitarkanika," wild.allindia.com, July 27, 1998.

3   Vandana Shiva, "Ecology and the Politics of Survival Conflicts over Natural Resources in India," New Delhi: SAGE Publications, 1991.

4   *Ecologist Asia,* Vol. 3, No. 4, July/August 1995.

5   United Nations Food and Agricultural Organization (FAO), "The State of the World Fisheries and Agriculture," Rome, 1995.

6   Vandana Shiva, "Ecology and the Politics of Survival," New Delhi: SAGE Publications, 1991, p. 320.

7   Tim Eichenberg and Durwood Zaelke, "Amicus Submission in the United States—Import Prohibition on Certain Shrimp and Shrimp Products Dispute," Washington, DC: CIEL, 1997, pp. 18–19.

8   N. Suresh, "Aquaculture is Answer to Fish Shortage," *The Times of India,* May 5, 1996.

9   FAO, Fisheries Department, "Agriculture, Towards 2010," Rome: November 1993, p. 183.

10   Vandana Shiva, "Globalization of Agriculture and the Growth of Food Insecurity," New Delhi: Research Foundation for Science, Technology, and Ecology (RFSTE), 1996.

11   FAO, "Review of the World Fishery Resources: Aquaculture," FAO Fisheries Circular, 886, 1995, pp. 1–127.

12   Steve Creech, "Sweet 'n' Sour Prawns—Shrimping in South East Asia," *Appropriate Technology,* Vol. 22, No. 2, September 1995, p. 25.

13   "Farming Has Expanded on World Bank Millions," *Fish Farming International,* July 1994, p. 10.

14   World Bank, *India Shrimp and Fish Culture,* Washington, DC: December 1991, p. 1.

15   Vandana Shiva and Gurpreet Karir, "*Chenmmeenkettu*: Towards Sustainable Aquaculture," New Delhi: RFSTE, 1996, p. 16.

16   These subsidies included up to 25 percent of capital investment or Rs. 30,000 per hectare up to a maximum of Rs. 150,000 for new farm development; a 25 percent subsidy up to a maximum of Rs. 500,000 for the establishment of medium-scale shrimp hatcheries of 30 million seed/year capacity and above; a 25 percent subsidy for feed and seed up to Rs. 3,000 and Rs. 450 per hectare, respectively; a 25 percent subsidy for the establishment of broodstock bank up to a maximum of Rs. 150,000; and finally, shrimp farmers are allowed to import shrimp feed at concessional customs duty.

17   Solon Barraclough and Andrea Finger-Stitch, "Some Ecological and Social Implications of Commercial Shrimp Farming in India," Geneva: UNRISD and WWI, March 1996, p. 31.

18   Jamulur Rahman and Frederick Vande Vusse, "Mangrove Forests: A Valuable But Threatened Indo-Pacific," Agriculture Department of the Asian Development Bank, p. 9.

19   S. Liyanage, "Pilot project on participatory management of Seguwanthive mangrove habitat in Puttalan District of Sri Lanka," International Conference on Wetlands and Development, Selangor, Malaysia, October 8–14, 1995.

20   M.S. Tuan, "Proceedings of Ecotone v. Community Participation in Conservation, Sustainable Use and Rehabilitation of Mangroves in Southeast Asia," eds. P.N. Hong et al., UNESCO, 1997.

21   A. Alvarez, B. Vasconez, and L. Guerrero, "Establishing a Sustainable Shrimp Mariculture Industry in Ecuador," eds. S. Olsen and L. Arriaga, *Multi-temporal study of mangrove, shrimp farm and salt flat areas in the coastal zone of Ecuador, through information provided by remote sensing.* University of Rhode Island Coastal Resources Center, USA; Ministerio de Energia y Minas, Ecuador; and US Agency for International Development, USA; 1989, pp. 141–46.

22   P. Menasveta, "Mangrove destruction and shrimp culture systems," *Asian Shrimp News,* 1996.

23   Justice Suresh et al., "Expert Committee Report on Impact of Shrimp Farms along the Coast of Tamil Nadu and Pondicherry," submission to the Supreme Court of India, 1995, p. 37.

24   FAO, "Reducing Environmental Impact of Coastal Aquaculture," FAO Reports and Studies, 47, p. 5.

26   Voices from the Grassroots, national public hearing, National Solidarity Convention on Aquaculture and Protection of the Coastal Zone, New Delhi, India, July 10, 1997.

25   RFSTE, "Facts about the Shrimp Industry," based on National Public Hearing and PUCL Report, July 10, 1997.

26   Statements made at National Public Hearing on the Aquaculture Industry, organized by PUCL and RFSTE, New Delhi, July 10, 1997.

27   Karol Wrage, "Biogrow Salmon receives grant," *Biotech Reporter,* April 1995, p. 7.

28   Anne Kapuscinski and Eric Hallerman, "Transgenic Fish and Public Policy: Anticipating Environmental Impacts of Transgenic Fish," *Fisheries,* Vol. 15, No. 1, p. 5.

29   "Native Fish, Introduced Fish: Genetic Implications," National Audobon Society, 1992.

MAD COWS
and SACRED COWS

4

W hen I gave a speech at the Dalai Lama's 60th birthday celebration, he wrote me two beautiful lines of compassion: "All sentient beings, including the small insects, cherish themselves. All have the right to overcome suffering and achieve happiness. I therefore pray that we show love and compassion to all."[1]

What is our responsibility to other species? Do the boundaries between species have integrity? Or are these boundaries mere constructs that should be broken for human convenience? The call to "transgress boundaries" advocated by both patriarchal capitalists and postmodern feminists cannot be so simple. It needs to be based on a sophisticated and complex discrimination between different kinds of boundaries, an understanding of whom is protected by what boundaries and whose freedom is achieved by what transgressions.

In India, cows have been treated as sacred—as Lakshmi, the goddess of wealth, and as the cosmos in which all gods and goddesses reside—for centuries. Ecologically, the cow has been central to Indian civilization. Both materially and conceptually the world of Indian agriculture has built its sustainability on the integrity of the cow, considering her inviolable and sacred, seeing her as the mother of the prosperity of food systems.

According to K.M. Munshi, India's first agriculture minister after independence from the British, cows

> are not worshipped in vain. They are the primeval agents who enrich the soil—nature's great land transformers—who supply organic matter which, after treatment, becomes nutrient matter of the greatest importance. In India, tradition, religious sentiment, and economic needs have tried to maintain a cattle population large enough to maintain the cycle.[2]

By using crop wastes and uncultivated land, indigenous cattle do not compete with humans for food; rather, they provide organic fertilizer for fields and thus enhance food productivity. Within the sacredness of the cow lie this ecological rationale and conservation imperative. The cow is a source of cow-dung energy, nutrition, and leather, and its contribution is linked to the work of women in feeding and milking cows, collecting cow dung, and nurturing sick cows to health. Along with being the primary experts in animal husbandry, women are also the food processors in the traditional dairy industry, making curds, butter, ghee, and buttermilk.

Indian cattle provide more food than they consume, in contrast to those of the U.S. cattle industry, in which cattle consume six times more food than they provide.[3] In addition, every year, Indian cattle excrete 700 million tons of recoverable manure: half of this is used as fuel, liberating the thermal equivalent of 27 million tons of kerosene, 35 million tons of coal, or 68 million tons of wood, all of which are scarce resources in India. The remaining half is used as fertilizer.

Two-thirds of the power requirements of Indian villages are met by cattle-dung fuel from some 80 million cattle. (Seventy million of these cattle are the male progeny of what industrial developers term "useless" low-milk-yielding cows.) To replace animal power in agriculture, India would have to spend about $1 billion annually on gas. As for other livestock produce, it may be sufficient to mention that the export of hides, skins, and other products brings in $150 million annually.[4]

Yet this highly efficient food system, based on multiple uses of cattle, has been dismantled in the name of efficiency and development. The Green Revolution shifted agriculture's fertilizer base from renewable organic inputs to non-renewable chemical ones, making both cattle and women's work with cattle dispensable in the production of food grain. The White Revolution, aping the West's wasteful animal husbandry and dairying practices, is destroying the world's most evolved dairy culture and displacing women from their role in the dairy-processing industry.

The Green Revolution has emerged as an enemy to the White, as the high-yielding crop varieties have reduced straw production, and their byproducts are unpalatable to livestock and thus useless as fodder. Further, hybrid crops deprive the soil of nutrients, creating deficiencies in fodder and disease in livestock. The White Revolution, in turn, instead of viewing livestock as ecologically integrated with crops, has reduced the cow to a mere milk machine. As Shanti George observes,

> The trouble is that when dairy planners look at the cow, they see just her udder; though there is much more to her. They equate cattle only with milk, and do not consider other livestock produce—draught power, dung for fertilizer and fuel, hides, skins, horn, and hooves.[5]

In India, cow's milk is but one of the many byproducts of the interdependence between agriculture and animal husbandry. There, cattle are considered agents of production in the food system; only secondarily are they viewed as producing consumable items. But the White Revolution makes milk production primary and exclusive, and according to the Royal Commission and the Indian Council of Agricultural Research, if milk production is unduly pushed up, it may indirectly affect the entire basis of Indian agriculture.[6]

Worse, trade-liberalization policies in India are leading to the slaughter of cattle for meat exports, threatening diverse, disease-resistant breeds and small farmers' integrated livestock-crop-production systems with extinction. In the United Kingdom, giant slaughterhouses and the factory farming of cattle are being called into question by the spread of "mad cow disease" (BSE—bovine spongiform encephalopathy), which has infected over 1.5 million cows in Britain. While this disease is sounding the death knell of the non-sustainable livestock economy in Britain, India's "sacred cows" are being sent to slaughterhouses to "catch up" with the beef exports and beef consumption figures of "advanced" countries. This globalization of non-sustainable and hazardous systems of food production is symptomatic of a deeper madness than that infecting U.K. cows.

RATCHETING UP THE MILK MACHINE

*A*s the idea of the cow-as-milk-machine runs into trouble worldwide, multinational biotech industries are promoting new miracles of genetic engineering to increase milk production, further threatening the livelihoods of small producers. Multinational corporations such as Elanco (a subsidiary of Eli Lilly), Cynamic, Monsanto, and Upjohn are all rushing to put bovine somatrophin (BST), a growth hormone commercially produced by genetic engineering, on the market, in spite of controversy about its ecological impact.[7]

When injected daily into cows, BST diverts energy to milk production. Cows may get emaciated if too much energy is diverted to produce milk. And, as in all other "miracles" of modern agricultural science, the gain in milk production is contingent upon a number of other factors, such as use of industrial feed and a computerized feeding program.[8] Finally, women's traditional role in caring for cows and processing milk falls into the hands of men and machines.

The use of genetically engineered BST, or bovine growth hormone (BGH), is leading to major consumer resistance and a demand for the labeling of milk, which the biotechnology industry actively opposes. The European Union has voted against the labeling of genetically engineered products, and Monsanto has sued U.S. farmers who label their milk "BGH-free." Democracy is thus stifled by "free trade."

The inherent violence of the White Revolution lies in its treatment of the needs of small farmers and of living resources as dispensable if they produce the wrong thing in the wrong quantity. The same global commoditization processes that render Indian cattle "unproductive" (even when, considered holistically, they are highly productive) simultaneously dispense with European cattle for being overproductive. Annihilating diverse livestock destroys knowledge on how to protect and conserve living resources as sources of life. This protection is replaced by the protection of the profits of rich farmers and the control of agribusiness.

CROPS AS FOOD FOR ALL

*I*n ecological agricultural cultures, technologies and economies are based on an integration between crops and animal husbandry. The wastes of one provide nutrition for the other, in mutual and reciprocal ways. Crop byproducts feed cattle, and cattle waste feeds the soils that nourish the crops. Crops do not just yield grain, they also yield straw, which provides fodder and organic matter. Crops are thus food for humans, animals, and the many organisms in the soil. These organically fed soils are home to millions of microorganisms that work and improve the soil's fertility. Bacteria feed on the cellulose fibers of straw that farmers return to the soil. In each hectare, between 100 and 300 kilograms of amoebas feed on these bacteria, making the lignite fibers available for uptake by plants. In each gram of soil, 100,000 algae provide organic matter and serve as vital nitrogen fixers. In each hectare are one to two tons of fungi and macrofauna such as anthropods, mollusks, and field mice. Rodents that bore under the fields aerate the soil and improve its water-holding capacity. Spiders, centipedes, and insects grind organic matter from the surface of the soil and leave behind enriching droppings.[9]

Soils treated with farmyard manure have from 2 to 2.5 times as many earthworms as untreated soils. These earthworms contribute to soil fertility by maintaining soil structure, aeration, and drainage and by breaking down organic matter and incorporating it into the soil. According to Charles Darwin, "It may be doubted whether there are many other animals which have played so important a part in the history of creatures."[10]

The little earthworm working invisibly in the soil is actually a tractor, fertilizer factory, and dam combined. Worm-worked soils are more water-stable than unworked soils, and worm-inhabited soils have considerably more organic carbons and nitrogen. By their continuous movement through soils, earthworms aerate the soil, increasing the air volume in soil by up to 30 percent. Soils with earthworms drain four to ten times faster than soils without earthworms, and their water-holding capacity is 20 percent higher. Earthworm casts, or droppings, which can consist of up to 36 tons per acre per year, contain carbon, nitrogen, calcium, magnesium, potassium, sodium, and phosphorous, promoting the microbial activity essential to soil fertility.

Industrial-farming techniques would deprive these diverse species of food sources and instead assault them with chemicals, destroying the rich biodiversity in the soil and with it the basis for the renewal of soil fertility.

THE INTENSIVE LIVESTOCK ECONOMY

Europe's intensive livestock economy requires seven times the area of Europe in other countries for the production of cattle feed.[11] These "shadow acres" necessary for feed production are in fact an extensive use of resources. While this feed-production system does not conserve acres, the concentration of animals in unlivable spaces does save space. The efficiency question that the intensive livestock industry is always asking is, "How many animals can be crammed into the smallest space for the least cost and the greatest profit?"[12]

In a complementary system of agriculture, the cattle eat what the humans cannot. They eat straw from the crops and grass from pastures and field boundaries. In a competitive model such as the livestock industry, grain is diverted from human consumption to intensive feed for livestock. It takes two kilograms of grain to produce one kilogram of poultry, four kilograms of grain to produce one kilogram of pork, and eight kilograms of grain to produce one kilogram of beef.

Cows are basically herbivores. The biomass they eat is digested in the rumen, the huge first chamber of the four stomachs of the cow. The livestock industry has increased cows' milk and meat production by giving them intensive, high-protein feed concentrate, an inappropriate diet since cows need roughage. One of the methods developed by the livestock industry to circumvent this need for roughage is by feeding them plastic pot-scrubbing pads. The scrubbing pads remain in the rumen for life.[13]

Robbing cattle of the roughage they need does not merely treat them unethically; it also does not reduce the acreage needed to feed the cows, since the concentrate comes from grain that could have fed people. The shift from a cooperative, integrated system to a competitive, fragmented one creates additional pressures on scarce land and grain resources. This in turn leads to non-sustainability, violence to animals, and lower productivity when all systems are assessed.

BREAKING BOUNDARIES: TRANSFORMING HERBIVORES INTO CANNIBALS

*A*s food for animals from farms disappears, animal feed is based increasingly on other sources, including the carcasses of dead animals. This is how the conditions for the mad-cow-disease epidemic were created. BSE infection, known as "scrapie" in sheep, typically bores into the brain and the nervous system and does not show itself as a disease until the infected animals are adults. Infected cows are nervous and shaky, and rapidly descend into dementia and death. Dissection of affected cows shows that their brains have disintegrated and are full of holes. In humans, this disease is called Creutzfeldt-Jakob disease, named after two German doctors.

The first case of BSE in the United Kingdom was confirmed in November 1986. By 1988, more than 2,000 cases of BSE had been confirmed. By August 1994, there were 137,000 confirmed cases, more than six times the number predicted by the government in their "worst case scenario."

The epidemic spread by feeding healthy cattle the remains of infected cattle. In 1987, 1.3 million tons of animal carcasses were processed into animal feed by "rendering plants." The largest portion of the animal material processed, 45 percent, came from cows. Pigs contributed 21 percent, poultry 19 percent, and sheep 15 percent. This created 350,000 tons of meat and bone meal and 230,000 tons of tallow.[14] Sheep infected with scrapie were thus fed to cows, which contracted BSE, and their carcasses were again fed to cattle. By 1996, more than 1.6 million cattle had become victims of BSE.

British farmers, increasingly dependent on industrial cattle feed, demanded that the sources of cattle feed be labeled, but the feed industry has denied farmers' and consumers' "right to know." Instead, the feed industry has been labeling its feed on the basis of its chemical constitution, thus camouflaging its biological sources.

THE BSE EPIDEMIC: CROSSING SPECIES BARRIERS

*W*hen the BSE epidemic broke out, scientists started to warn that if the disease had jumped from sheep to cows, there was every possibility that it could shift from cows to humans. The government continued to state this was impossible.

But in January 1996, a degenerative brain disorder in ten children was linked to the consumption of beef infected with BSE. Ten thousand schools stopped serving beef in their meals. Many countries in Europe and as far away as New Zealand and Singapore have stopped importing U.K. beef. In April 1996, the European Union announced that it would help fund the mass slaughter of 4.7 million British cattle.[15]

By repeatedly denying the method of BSE transmission, by refusing to call for the biological labeling of animal feed, and by other evasions, both the government and official scientists colluded in exacerbating the BSE epidemic. In an economy in which trade is not subjected to ethical, ecological, and health imperatives, "science" that serves commerce will systematically mislead citizens. Even as new diseases threaten the lives and health of farm animals and consumers, official scientific agencies keep repeating the mantra of "no hard scientific evidence." In the meantime, consumers are making their own decisions, voting against hazardous factory farming by boycotting beef.

European consumption of U.K. beef and beef products dropped by 40 percent, and the European Union was forced to ban the export of U.K. beef and beef products.

THE NEW APARTHEID: CONTAMINATED BEEF FOR THE SOUTH

*I*n 1991 the chief economist of the World Bank suggested that, because people are poorer and life is cheaper in the Third World, exporting toxics there made economic sense. In an internal memo, Lawrence Summers wrote,

> Just between you and me, shouldn't the World Bank be encouraging more migration of the dirty industries to the LDCs [less developed countries]?... The economic logic behind dumping a load of toxic waste in the lowest wage country is impeccable, and we should face up to that.... Under-populated countries in Africa are vastly under-polluted; their air quality is probably vastly inefficiently low compared to Los Angeles or Mexico City.... The concern over an agent that causes a one-in-a-million change in the odds of prostate cancer is obviously going to be much higher in a country where people survive to get prostate cancer than in a country where under-five mortality is 200 per thousand.[16]

In these economics of genocide, largely white, male elites of the North create class, race, and gender boundaries to exclude other social groups from the fundamental human rights to life and safety. This blatant disregard for the rights of Third World people was reinforced in 1996, when the European Union lifted its ban on the export of possibly BSE-infected U.K. beef and bovine products for Third World countries.

There is a difference between ecological boundaries and socially constructed boundaries. The difference between herbivores and carnivores is an ecological boundary. It needs to be respected for the sake of both cows and humans. The difference between the value of human life in the North and South is a politically constructed boundary. It needs to be broken for the sake of human dignity.

TRANSFORMING VEGETARIANS
INTO BEEF-EATERS

*A*t a time when meat consumption is declining in Western countries, India's trade-liberalization program is trying to convert a predominantly vegetarian society into a beef-eating one. This program is based on the false equation that the only source of protein is animal protein, and that higher animal consumption equals a higher quality of life.

According to Dr. Panya Chotiawan, chair of a Thai poultry producer, "protein...provides both strength and brain structure. Therefore, consuming sufficient protein will generate a healthier body and promote intelligence."[17]

However, it is not the case that higher animal-protein consumption makes for a better quality of life or higher intelligence. The trend is that people seeking a genuinely high quality of life are shifting to vegetarianism. In the United States, animal protein consumption has dropped, and the mad-cow-disease epidemic has also triggered people to move to vegetarianism.

Indians who are predominantly vegetarian are not unintelligent. Our source of protein is plant-based. Our diet has a rich variety of legumes, which provide healthy proteins for human consumption and a free enrichment of nitrogen for the soils. Most indigenous farming systems are based on polycultures, which include leguminous crops.

The three most important diseases of the affluent countries–cancer, stroke, and heart disease—are linked conclusively to consumption of beef and other animal products. International studies comparing diets in different countries have shown that diets high in meat result in more deaths from intestinal cancer per capita. Japanese people in the United States eating a high-meat diet are three times as likely to contract colon cancer as the those eating the Japanese low-meat diet.[18] Modern, intensive systems of meat production have exacerbated the health hazards posed by meat consumption. Modern meats have seven times more fat than non-industrial meats, as well as drug and antibiotic residues.

SLAUGHTERING INDIA'S CATTLE FOR EXPORTS

*T*he cultural attitudes that maintain the widespread vegetarianism in India are seen as obstacles to overcome in order to institute a new meat-eating culture. According to India's "New Livestock Policy,"

> The beef production in India is purely an adjunct to milk and draught power production. The animals slaughtered are the old and the infirm and the sterile, and are in all cases malnourished. There is no organized marketing and no grading system, and beef prices are at a level which makes feeding uneconomic. There is no instance of feedlots or even individual animals being raised for meat. Religious sentiments (particularly in the Northern and western parts of India) against cattle slaughter seem to spill over also on buffaloes and prevent the utilization of a large number of surplus male calves.[19]

The Ministry of Agriculture provides 100 percent grants and tax incentives to encourage the setting up of slaughterhouses. According to a 1996 Union Ministry of Environment report, at least 32,000 illegal slaughterhouses established themselves in the preceding five years. By 1995, the total quantity of meat exports had risen more than 20-fold, to 137,334 tons.[20] Total meat exports, including beef, veal, and buffalo meat, almost doubled between 1990 and 1995. But between 1991 and 1996, cattle, buffalo, and other livestock populations have only increased by half that rate. In other words, India is exporting more meat than is being replenished.

Meat exports are leading to a decline not only in livestock numbers, but also in the rich diversity of cattle breeds known for their hardiness, milk production, and draught power. According to the United Nations Food and Agriculture Organization, "the diversity of domestic animal breeds is dwindling rapidly. Each variety that is lost takes with it irreplaceable genetic traits–traits that may hold the key to resisting disease or to productivity and survival under adverse conditions."[21] If measures to arrest these trends are not taken now, most of us will witness the extinction of livestock within our very lifetimes, and with it the foundation of sustainable agriculture will disappear.

Another significant factor contributing to the decline of cattle is the shortage of fodder, stemming from the emphasis on grains bred for high yields, the planting of monocultures of non-fodder species such as eucalyptus, and the growing scarcity of grazing lands and pastures due to the enclosure of the commons.

The decline of animal wealth is destroying the rural economy and rural livelihoods. This will adversely affect the landless, the lowest castes, and women. Women provide nearly 90 percent of all labor for livestock management. Of the 70 million households that depend on livestock for their livelihoods, two-thirds are small and marginal farmers and landless laborers. Because of increased cattle exports, the price of livestock has escalated, and there is less and less dung available for manure and cooking fuel. More fertilizers, fossil fuels, tractors, and trucks must be imported to replace the energy and fertility that cattle gave freely to the rural economy. Thus, while animal exports are earning the country Rs. 10 million, the destruction of animal wealth is costing the country Rs. 150 million.

A case in point is one of the biggest export-oriented slaughterhouses, Al-Kabeer in Andhra Pradesh. Al-Kabeer slaughters 182,400 buffaloes every year, animals whose dung could have provided for the fuel needs of over 90,000 average Indian families of five. The government's transport of kerosene to replace this fuel costs hundreds of millions of rupees, which means that poor people pay vastly higher fuel expenses. In 1987–88, Rs. 5.5 billion of kerosene was imported. By 1992–93, this amount had increased almost fourfold.

If livestock were not slaughtered in the state of Andhra Pradesh, farmyard manure would cultivate 384 hectares, producing 530,000 tons of food grain.[22] The state of Andhra Pradesh must now spend Rs. 9.1 billion to import nitrogren, phosphorus, and potash previously provided by livestock over the duration of their lives. This means that against a projected earning of Rs. 200 million by Al-Kabeer through the killings, the state could actually save Rs. 9.1 billion in foreign exchange by not killing.[23]

Al-Kabeer has provided just 300 jobs. In contrast, small-scale slaughtering for local consumption creates livelihoods and allows all parts of an animal to be used. The skin is used for leather, and bones

and horns provide material for crafts and fertilizer. In large-scale industrial slaughterhouses, all these byproducts are treated as waste and
become a source of pollution. The entire area around Al-Kabeer is contaminated with blood, skin, and bones from slaughtered cattle.
Al-Kabeer has proposed to build a "rendering" plant to use this animal
waste to make cattle feed, yet another symptom of the mad-cow culture
replacing the sacred-cow culture.

In one lawsuit against Al-Kabeer, the court ordered a 50 percent reduction of its capacity in order to save the cattle wealth and the rural
economy of Andhra Pradesh. In another case involving a slaughterhouse, the judge ruled that instead of exporting meat, India should export a message of compassion. According to the judgement,

> This fundamental Duty in the Constitution to have compassion for
> all living creatures thus determines the legal relation between In-
> dian Citizens and animals on Indian soil, whether small ones or
> large ones.... Their place in the Constitutional Law of the land is
> thus a fountainhead of total rule of law for the protection of ani-
> mals and provides not only against their ill treatment, but from it
> also springs a *right to life* in harmony with human beings.
>
> If this enforceable obligation of State is understood, certain results
> will follow. *First*, the *Indian state cannot export live animals for
> killing;* and *second*, cannot become a party to the killing of animals
> by sanctioning exports in the casings and cans stuffed with dead
> animals after slaughter. Avoidance of this is preserving the Indian
> Cultural Heritage.... *India can only export a message of compas-
> sion towards all living creatures of the world,* as a beacon to pre-
> serve ecology, which is the true and common Dharma for all
> civilizations.[24]

But the Indian Constitution's protection of animals and rural livelihoods is being challenged by international trade agreements. In March
1998, the World Trade Organization announced the initiation of a dispute by the European Union (EU) against India's restriction on the export of raw hides and fur. The EU argues that preventing the free export
of furs and hides contravenes Article XI of the General Agreement on
Tariffs and Trade (GATT).[25] According to Article XI of GATT, any restriction on imports and exports is illegal, even though such restrictions
might be necessary for cultural, ecological, and economic reasons.[26]

Exporting raw hides and furs would threaten India's cattle wealth as well as the livelihoods of craftspeople, shoemakers, cobblers, farmers, and other small producers. In 1993, when India was forced to remove export restrictions on cotton, 2 million weavers lost their livelihoods.

MCDONALDIZATION

*G*lobalization has created the McDonaldization of world food, resulting in the destruction of sustainable food systems. It attempts to create a uniform food culture of hamburgers. The mad-cow-disease epidemic tells us something of the costs hidden in this food culture and food economy.

In 1994, Pepsi Food, Ltd., was given permision to start 60 restaurants in India: 30 each of Kentucky Fried Chicken (KFC) and Pizza Hut. The processed meats and chicken offered at these restaurants have been identified by the U.S. Senate as sources of the cancers that one American contracts every seven seconds. The chicken, which would come from an Indian firm called Venky's, would be fed on a "modern" diet of antibiotics and other drugs, such as arsenic compounds, sulfa drugs, hormones, dyes, and nitrofurans. Still, many chickens are riddled with disease, in particular chicken cancer (leukosis). They can also carry salmonellosis, which does not die with ordinary cooking.

Both KFC and Pizza Hut have guaranteed that they will generate employment. However, according to studies conducted by the Ministry of Environment on other meat industries, Al-Kabeer has displaced 300,000 people from their jobs, while employing only 300 people at salaries ranging from Rs. 500 to Rs. 2,000 per month. Venky's chicken has not employed one extra person after getting the contract for chicken supply from KFC and Pizza Hut. In fact, the company is being encouraged to mechanize further rather than use human labor.

Junk-food chains, including KFC and Pizza Hut, are under attack from major environmental groups in the United States and other developed countries because of their negative environmental impact. Intensive breeding of livestock and poultry for such restaurants leads to deforestation, land degradation, and contamination of water sources and other natural resources. For every pound of red meat, poultry, eggs, and milk produced, farm fields lose about five pounds of irreplaceable top soil. The water necessary for meat breeding comes to about 190 gallons per animal per day, or ten times what a normal Indian family is supposed to use in one day, if it gets water at all.

KFC and Pizza Hut insist that their chickens be fed on maize and soybean. It takes 2.8 kilograms of corn to produce one pound of chicken. Egg-layers also need 2.6 pounds of corn and soybean. Nearly seven pounds of corn and soybean are necessary to produce one pound of pork. Overall, animal farms use nearly 40 percent of the world's total grain production. In the United States, nearly 70 percent of grain production is fed to livestock.

Maize, though not a major food crop in India, has traditionally been grown for human consumption. Land will be diverted from production of food crops for humans to production of maize for chicken. Thirty-seven percent of the arable land in India will be diverted toward such production. Were all the grain produced consumed directly by humans, it would nourish five times as many people as it does after being converted into meat, milk, and eggs, according to the Council for Agricultural Science and Technology.

The food culture of India is as diverse as its ecosystems and its people, who use a variety of cereals, pulses, and vegetables as well as cooking methods to suit every need and condition. However, advertising is already having a negative impact on Indians' food and drink patterns. No longer are homemade snacks and lime juice or buttermilk offered to guests; instead, chips and aerated soft drinks are.

METAPHORS OF ECOLOGICAL CULTURE
AND INDUSTRIAL CULTURE

*T*he mad cow is a product of "border crossings" in industrial agriculture. It is a product of the border crossing between herbivores and carnivores. It is a product of the border crossing between ethical treatment of other beings and violent exploitation of animals to maximize profits and human greed.

Cross-breeding programs meant to "improve" Indian breeds with "superior" European breeds are resulting in cross-bred cattle, perceived only as milk machines. During the *Mattu Pongal* festival in India, villagers decorate, worship, and leave free to roam their livestock animals, but as far as I have seen, not their cross-bred cows. Meat export programs are converting the sacred cow into a meat machine, leading to a decline in livestock and eroding cattle diversity.

Species boundaries between humans and cattle are also being crossed to create pharmaceuticals in the milk of factory-farmed animals. This construction of "mammalian bioreactors" is the ultimate step in the reduction of cows to machines.

These border crossings, promoted by corporate elites for profit, are rationalized by the popular postmodern stances taken by some academics. As technofeminist Donna Haraway writes:

> Transgenic border crossing signifies serious challenges to the "sanctity of life" for many members of Western cultures, which historically have been obsessed with racial purity, categories authorized by nature, and the well-defined self.... In opposing the production of transgenic organisms, especially opposing their patenting and other forms of private commercial exploitation, committed activists appeal to notions such as the integrity of natural kinds and the natural types or self-defining purpose of all life forms.[27]

This academic rationale for an attack on environmental and Third World movements to safeguard their food and livelihoods is based on many false assumptions. The first is that the "sanctity of life" is merely a Western construct. Diverse cultures, animal-rights activists, and ecologists all believe in the need for respect for all living things. The sanctity of life is characteristic of the worldviews of diverse indigenous cultures. As Jerry Mander has indicated, Western industrial civilization has evolved in the absence of the sacred.[28]

The second flawed assumption is to equate "sanctity of life" with racism and an obsession with racial purity. In fact, racism and life's sanctity are mutually exclusive. The racist obsessed with "racial purity" indulges in "ethnic cleansing" and violates the sanctity of life. The existence of diversity and difference in itself does not lead to racism. It is when that diversity is hierarchially ordered on the basis of "superiority" that we get racism. Anti-racism does not require wiping out the blackness of the black or the brownness of the brown, it requires resisting the view that sees black and brown as inferior to white. In fact, during the apartheid regime of South Africa, "border crossing" between whites and Blacks did not create liberation for the Blacks, it created new oppression.[29]

A cow is not merely a milk machine or a meat machine, even if industry treats it in such a way. That is why cows are hurt by the industrial treatment they are subjected to. When forced to become carnivores instead of herbivores, they become infected with BSE. When injected with growth hormones, they become diseased. To deny subjecthood to cows and other animals, to treat them as mere raw material, is to converge with the approach of capitalist patriarchy.

Sacred cows are the symbols and constructions of a culture that sees the entire cosmos in a cow, and hence protects the cow to protect ecological relations as well as the cow as a living being, with its own intelligence and its own self-organizing capacity. Referring to the self-organized nature of animals and other living organisms, Goethe concluded,

> Hence we conceive of the individual animal as a small world, existing for its own sake, by its own means. Every creature has its own reasons to be. All its parts have a direct effect on one another, a relationship to one another, thereby consistently renewing the circle of life.[30]

Mad cows are symbols of a worldview that perceives no difference between machines and living beings, between herbivores and carnivores, or between the Sindhi and Sahiwal and the Jersey and the Holstein. Sacred cows are a metaphor of ecological civilization. Mad cows are a metaphor of an anti-ecological, industrial civilization.

At the threshhold of the third millennium, liberation strategies have to ensure that human freedom is not gained at the cost of other species, that freedom for one race or gender is not based on increased subjugation of other races and genders. In each of these strivings for freedom, the challenge is to include the Other.

For more than two centuries, patriarchal, eurocentric, and anthropocentric scientific discourse has treated women, other cultures, and other species as objects. Experts have been treated as the only legitimate knowers. For more than two decades, feminist movements, Third World and indigenous people's movements, and ecological and animal-rights movements have questioned this objectification and denial of subjecthood.

Ecological feminisms recognize the intrinsic worth of all species, the intelligence of all life, and the self-organizational capacity of beings. They also recognize that there is no justification in a hierarchy between knowledge and practice, theory and activism, academic thought and everyday life. Such hierarchies have no epistemological basis, though they do have a political basis. In this perspective, it is not just the Western industrial breeders whose knowledge counts and whose knowledge should displace all other knowers: indigenous cattle breeders, farmers, women, and animals.

REVERSING THE MCDONALDIZATION OF THE WORLD

"What man does to the web of life, he does to himself." How we relate to other species will determine whether the third millennium will be an era of disease and devastation, and of exclusion and violence, or rather a new era based on peace and non-violence, health and well-being, inclusiveness and compassion.

Unsustainable outcomes are the inevitable result of the deepening of patriarchal domination over ways of knowing and relating non-violently to what have been identified as "lesser species," including women. But sustainability can be created by an inclusive feminism, an ecological feminism, in which the freedom of every species is linked to the liberation of women, in which the tiniest life form is recognized as having intrinsic worth, integrity, and autonomy.

Women of our generation especially have to decide whether to protect the knowledge and wisdom of our grandmothers in the maintenance of life or whether to allow global corporations to push most species to extinction, mutilate and torture those that are found profitable, and undermine the health and well-being of the earth and its communities.

The mad cow, as a product of border crossings, is a "cyborg" in Donna Haraway's brand of "cyborg" feminism.[31] According to Haraway, "I'd rather be a cyborg than a goddess."[32] In India, the cow is Lakshmi, the goddess of wealth. Cow dung is worshipped as Lakshmi because it is the source of renewal of the earth's fertility through organic manuring. The cow is sacred because it is at the heart of the sustainability of an agrarian civilization. The cow as goddess and cosmos symbolizes care, compassion, sustainability, and equity.

From the point of view of both cows and people, I would rather be a sacred cow than a mad one.

1 Special issue on "Reverence for Life," *Quarterly Monitor,* No. 13, New Delhi: Research Foundation for Science, Technology and Natural Resource Policy.

2 K.M. Munshi, "Towards Land Transformation," Government of India, Ministry of Food and Agriculture, 1951.

3 In India, cattle use 29 percent of the organic matter, 22 percent of the energy, and 3 percent of the protein provided to them, in contrast to 9, 7, and 5 percent respectively in the United States' intensive cattle industry. Shanti George, *Operation Flood,* Delhi: Oxford University Press, 1985, p. 31.

4 Shanti George, p. 31.

5 Shanti George, p. 30.

6 Shanti George, p. 59.

7 "Buttercup Goes on Hormones," *The Economist,* May 9, 1987.

8 B. Kneen, "Biocow," *Ram's Horn: Newsletter of the Nutrition Policy Institute,* Toronto, Ontario, No. 40, May 1987.

9 Claude Bourguignon, Address at ARISE workshop, Auroville, India, April 1995.

10 Charles Darwin, "The Formation of Vegetable Mould through the Action of Worms with Observations on their Habits," London: Faber and Faber, 1927.

11 "Sustainable Europe," Friends of the Earth (International), 1995.

12 David Coats, *Old MacDonald's Factory Farm,* New York: Continuum, 1989, p. 73.

13 Trials indicated that steers fed 100 percent concentrate plus pot scrubbers grew at approximately the same rate as cattle fed 85 percent concentrate with 15 percent roughage. S. Loerch, "Efficiency of plastic pot scrubbers as a replacement for roughage in high concentrate cattle diets," *Journal of Animal Science,* No. 60, 1991, pp. 2321–28.

14 Richard W. Lacey, *Mad Cow Disease: The History of BSE in Britain,* Channel Islands: Cypsela Publications Limited, 1994, p. 32.

15 "EU agrees to fund slaughter of millions of British cattle," Cable News Network, April 3, 1996.

16 Lawrence Summers, quoted in Vandana Shiva, "Ecological Balance in an Era of Globalization," *Global Ethics and Environment,* ed. Nicholas Low, London: Routledge, 1999.

17 Panya Chotiawan, quoted in D. Juday, "Intensification of Agriculture and Free Trade," Paper presented at VIII World Conference on Animal Production, Seoul, Korea, June 28–July 4, 1998.

18 Vandana Shiva, "The New Livestock Policy: A Policy of Ecocide of Indigenous Cattle Breeds and a Policy of Genocide for India's Small Farmers," New Delhi: Research Foundation for Science, Technology, and Ecology, 1995.

19   "New Livestock Policy," Section 2.10 on "Meat Production," Ministry of Agriculture, Department of Animal Husbandry, 1995.

20   www.fao.org, 1996.

21   Some of the declining indigenous breeds today are Pangunur, Red Kandhari, Vechur, Bhngnari, Dhenani, Lohani, Rojhan, Bengal, Chittagong Red, Napalees Hill, Kachah, Siri, Tarai, Lulu and Sinhala. "The Hindu Survey of Indian Agricuture," *The Hindu,* 1996, p. 115.

22   Calculated on the basis of the average food grain produced per hectare in 1991, 1.382 tons.

23   The annual availability of major nutrients in farmyard manure, from the dung and urine of 1,924,000 buffaloes and 570,000 sheep per year works out to 11,171.79 tons of nitrogen, which at the current price of Rs. 20.97 per kg at unsubsidized rates, costs Rs. 234.2 million to import; 2,164.15 tons of phosphorus, which at the current price of Rs. 21.25 per kg. at unsubsidized rates, costs Rs. 46 million; 10,069.29 tons of potash, which at the current price of Rs. 8.33 per kg at unsubsidized rates, costs Rs. 83.9 million; for a total import cost of Rs. 364.1 million. Taking into account an average remaining life span of five years, the cost of importing goods previously produced by livestock equals Rs. 1.8 billion. Following the same argument, if all the animals which are going to be killed during five years of Al-Kabeer's operation live out their natural life spans, then the state would have to spend Rs. 9.1 billion on imports. In five years, Al-Kabeer has killed 920,000 buffaloes and 2,850,000 sheep to earn only Rs. 200 million a year, according to the company's own projections. It has provided just 300 jobs. Maneka Gandhi, "The Crimes of Al-Kabeer," *People for Animals Newsletter,* May 1995.

24   Tis Hazari Court, Judgement passed on March 23, 1992, Case No. 2267/90, Delhi.

25   Renato Ruggiero, speech given at "Policing the World Economy" Conference held at Geneva, March 23–25, 1998.

26   World Trade Organization, GATT Agreement, Geneva, 1994.

27   Donna Haraway, *Female Man* ©—*Meets—Onco Mouse*[TM], New York: Routledge, 1997, p. 80.

28   Jerry Mander, *In the Absence of the Sacred,* Sierra Club Books, 1995.

29   Some go so far as to suggest that gene transfer could "cure" racist attitudes in society. But on the contrary, "gene enhancement" therapy is being requested for changing skin color. (See Rick Weiss, "Gene Enhancements' Thorny Ethical Traits," *Washington Post,* October 11, 1997.) Genetic engineering is showing every sign of becoming the basis of a new racism, in which the blue-eyed, blond-haired, white-skinned race becomes the measure for all.

30   J.W. Goethe, *Scientific Studies,* ed. Douglas Miller, New York: Suhrkamp, p. 121.

31   Donna Haraway, "A Manifesto for Cyborgs: Science, Technology, and Socialist Feminism in the 1980s," *Socialist Review*, Vol. 80, pp. 65–108.

32   Donna Haraway, "A Manifesto for Cyborgs."

The STOLEN HARVEST
of SEED

$\mathcal{F}$or more than 10,000 years, farmers have worked with nature to evolve thousands of crop varieties to suit diverse climates and cultures. Indian farmers have evolved thousands of varieties of rice. Andean farmers have bred more than 3,000 varieties of potatoes. In Papua New Guinea, more than 5,000 varieties of sweet potatoes are cultivated.

This tremendous diversity has been the basis of our food supply, but today it is under threat from genetic erosion and genetic piracy. Monocultures and monopolies are destroying the rich harvest of seed given to us over millennia by nature and farming cultures.

From the 250,000 to 300,000 species of plants alive today, at least 10,000 to 50,000 are edible. Seven thousand species have been farmed and used for food. Just 30 species provide 90 percent of world calorie intake, and only four species—rice, maize, wheat, and soybean— provide most of the calories and proteins consumed by the world's population through global trade.

As Hope Shand of Rural Advancement Foundation International (RAFI) has stated,

> There is no doubt about the global economic importance of
> these major crops, but the tendency to focus on a small number
> of species masks the importance of plant species diversity to the
> world food supply. A very different picture would emerge if we
> were to look into women's cooking pots and if we could survey
> local markets and give attention to household use of non-
> domesticated species.[1]

Local markets and local cultures have allowed crop diversity to
thrive in our fields, enabling farmers to continue evolving diverse
breeds and conserving seeds and plant varieties. Ensuring the continued use of these seeds and plants is the best way to conserve them;
whichever economic system determines how plant species are used
also influences which species will survive and which will be pushed to
extinction.

As global markets replace local markets, monocultures replace diversity. Traditionally, 10,000 wheat varieties were grown in China.
These had been reduced to only 1,000 by the 1970s. Only 20 percent of
Mexico's maize diversity survives today. At one time, more than 7,000
varieties of apples were grown in the United States. More than 6,000
are now extinct. In the Philippines, where small peasants used to cultivate thousands of traditional rice varieties, just two Green Revolution
varieties occupied 98 percent of the entire rice-growing area by the
mid-1980s.

In 1996, the United Nations Food and Agriculture Organization
(FAO) organized the Leipzig Conference on Plant Genetic Resources,
which identified the introduction of new crop varieties as the single
most important cause of this massive loss of species diversity and native seeds. But diversity is under assault not just by monocultures but
also by monopolies.

MONOCULTURES AND MONOPOLIES

*I*ndustrial agriculture promotes the use of monocultures because of
its need for centralized control over the production and distribution of
food. In this way, monocultures and corporate monopolies reinforce
each other. Today, three processes are intensifying monopoly control

over seed, the first link in the food chain: economic concentration, patents and intellectual property rights, and genetic engineering.

Monsanto, which was earlier recognized primarily through its association with Agent Orange, today controls a large section of the seed industry. Between 1995 and 1998, Monsanto spent over $8 billion buying seed companies. Monsanto holds a controlling interest in Calgene, a California-based plant biotechnology firm that launched the "Flavr-Savr" tomato. In 1996, it bought the biotechnology assets of Agracetus, a subsidiary of W.R. Grace, for $150 million. In 1997, it purchased Asgrow from Seminis for $267 million.

In November 1997, Monsanto acquired Holden Seeds at 30 times its market value. Between 25 and 30 percent of the U.S. corn acreage is estimated to be planted with Holden seeds. In May 1998, Monsanto announced a $2.3 billion takeover of Dekalb, the United States's second-largest corn company, making Monsanto the dominant player in the corn market.

For $1.8 billion, Monsanto purchased Delta and Pine Land, giving Monsanto an overwhelming 85 percent share of the U.S. cottonseed market and a dominant global position in the cotton farming industry. Monsanto also now owns the joint U.S. Department of Agriculture (USDA)-Delta and Pine Land patent for what's been called "terminator technology," a method of creating sterile seeds.

In July 1998, Monsanto bought Unilever's European wheat-breeding business for $525 million. This acquisition is part of its push to monopolize the production and sale of genetically engineered wheat. Monsanto has also bought a large stake in India's largest seed company, MAHYCO, at 24 times the market value, and has formed a Monsanto-MAHYCO joint venture. According to Monsanto's Jack Kennedy, the company plans to "penetrate the Indian agricultural sector in a big way. MAHYCO is a good vehicle."[2] For $1.4 billion, Monsanto bought Cargill's international seed operations in Central and Latin America, Europe, Asia, and Africa.

Dominating the seed, pesticide, food, pharmaceutical, and veterinary products industries along with Monsanto are Novartis, which was formed via a merger of Sandoz and Ciba-Geigy, and Aventis, which was formed with the merger of Astra/Zeneca and DuPont. DuPont has fully acquired Pioneer Hi-bred, the world's largest seed company, which, according to *The Wall Street Journal*, "effectively divides most of the U.S. seed industry between DuPont and Monsanto."[3]

THE TERMINATOR LOGIC: ENGINEERING TOTAL CONTROL

*I*n March 1998, the USDA and the Delta and Pine Land Company announced the joint development and patent on a new agricultural biotechnology benignly called "Control of Plant Gene Expression." The new patent permits its owners and licensees to create sterile seeds by selectively programming the plant's DNA to kill its own embryos. The patent, which has been applied for in at least 78 countries, applies to plants and seeds of all species. The USDA, a government agency, receives a 5 percent profit from the sales of these seeds, which it considers a built-in "gene police."[4]

The result? If farmers save the seeds of these plants at harvest for future crops, the next generation of plants will not grow. Pea pods, tomatoes, peppers, heads of wheat, and ears of corn will essentially become seed morgues. Thus the system will force farmers to buy new seeds from seed companies every year. RAFI and other groups have dubbed the method "terminator technology," claiming that it threatens farmers' independence and the food security of over 1 billion poor farmers in Third World countries.

According to USDA scientist Melvin Oliver,

> The need was there to come up with a system that allowed you to self-police your technology, other than trying to put on laws and legal barriers to farmers saving seed, and to try and stop foreign interests from stealing the technology.[5]

Molecular biologists are currently examining the risk of the terminator function escaping the genome of the crops into which it has been intentionally incorporated and moving into surrounding open-pollinated crops or wild, related plants in nearby fields. Given nature's incredible adaptability and the fact that the technology has never been tested on a large scale, the possibility that the terminator may spread to surrounding food crops or to the natural environment is a serious one. The gradual spread of sterility in seeding plants would result in a global catastrophe that could eventually wipe out higher life forms, including humans, from the planet.

According to RAFI, "if the Terminator Technology is widely utilized, it will give the multinational seed and agrochemical industry an unprecedented and extremely dangerous capacity to control the world's food supply."[6] By RAFI's estimate, by 2010 the terminator and related-seeds market could constitute 80 percent or more of the entire global commercial-seed market, valued at $20 billion per year.

Third World governments and farmers have rejected these "gene control" technologies. The Indian government has stated that it will not allow the terminator technology to enter India. The Consultative Group on International Agricultural Research, the world's most important agricultural research system, has stated firmly that it will not use the technology in its breeding work. In response to Monsanto's planned advertising campaign "Let the Harvest Begin," African governments wrote a declaration "Let the Harvest Continue!" in which they stated,

> We do not believe that such companies or gene technologies will help our farmers to produce the food that is needed in the 21st century. On the contrary, we think they will destroy the diversity, the local knowledge, and the sustainable agricultural system that our farmers have developed for millennia, and that they will thus undermine our capacity to feed ourselves.[7]

According to writer Geri Guidetti,

> Never before has man created such an insidiously dangerous, far-reaching, and potentially "perfect" plan to control the livelihoods, food supply, and even survival of all humans on the planet. In one broad, brazen stroke of his hand, man will have irretrievably broken the plant-to-seed-to-plant-to-seed cycle, the cycle that supports most life on the planet. No seed, no food, unless you buy more seed. The Terminator Technology is brilliant science and arguably "good business," but it has crossed the line, the tenuous line between genius and insanity. It is a dangerous, bad idea that should be banned. Period.[8]

When Third World farmers sow seed, they pray, "May this seed be exhaustless." Monsanto and the USDA, on the other hand, seem to be saying, "Let this seed be terminated so that our profits and monopoly will be exhaustless."

Corporations argue that such technology is necessary in order for them to recoup their investment. But this argument would mean that arms manufacturers must be allowed to sell arms and that the nuclear industry should be freely allowed to make bombs. As humans with a duty to preserve life on the planet, we have a duty to stop certain activities on social and ecological grounds no matter how profitable they may be.

As a result of international outrage, Monsanto announced in October 1999 that it would abandon its plans to commercialize terminator technology. However, Monsanto will continue to develop other hazardous technologies, including those to control seed.[9]

SEED PIRACY

Seed and crops have been celebrated as sources of life's renewal and as the embodiment of fertility. In Asia, rice has been an important source of both nourishment and cultural identity.

Rice evolved as a food source in Asia. Asian Rice, *Oryza sativa,* has two subspecies, *indica* and *japonica.* The *japonica* varieties are shorter, rounder, and more translucent, while the *indica* varieties have longer, more slender grains that stay separate when cooked.

In Japan, rice and rice paddies are important as metaphors of "self." According to Emike Ohnuiki-Trerney, author of *Rice as Self,* "Agrarian rituals enact a cosmic cycle of gift exchange during which a new crop of rice is offered in return for the original seeds given by the deities."[10]

In India, rice is identified with *prana,* or life breath. Before the Green Revolution introduced monocultures that destroyed species diversity, more than 200,000 varieties of rice were grown in India. These indigenous rice varieties had evolved to survive floods and droughts, to thrive in uplands and coastal ecosystems, and to offer enhanced taste and medicinal value.

On the Indian subcontinent, Basmati rice has been grown for centuries and is referred to in ancient texts, folklore, and poetry.[11] This naturally perfumed variety of rice has always been treasured and eagerly coveted by foreigners.

Years of research on Basmati strains by Indian and Pakistani farmers have resulted in a diverse range of Basmati varieties. Their superior qualities are a result of these farmers' informal breeding and innovation. Today, there are 27 distinct, documented varieties of Basmati grown in India. A native-seed conservation program, Navdanya, has saved, collected, and distributed 14 Basmati varieties.

In recent years, Basmati rice has been one of India's fastest-growing export items. Every year, India grows 650,000 tons of Basmati, covering 10 to 15 percent of the total land area under rice cultivation in India. Annually, between 400,000 and 500,000 tons of Basmati are exported. The main importers of Indian Basmati are the Middle East (65 percent), Europe (20 percent), and the United States (10 to 15 percent). At $850 a ton, Indian Basmati is the most expensive rice being imported by the European Union. Pakistani Basmati costs $700 a ton, and Thai fragrant rice costs $500 a ton.[12]

A recent patent, however, threatens to pirate farmers' innovation, and monopolizes this trade. On September 2, 1997, the Texas-based RiceTec, Inc. was granted patent number 5663484 on Basmati rice lines and grains. RiceTec got patent rights on Basmati rice and grains while already trading the rice in its brand names such as Kasmati, Texmati, and Jasmati. The patent will allow RiceTec to sell internationally what it claims to be a new variety of Basmati, developed under the name of Basmati.

RiceTec's patented Basmati variety was derived from Indian Basmati crossed with semi-dwarf varieties including *indica* varieties. These varieties are farmers' varieties bred over centuries on the Indian subcontinent. RiceTec's method of crossing different varieties to mix traits–in this case, the Basmati characteristics from Basmati and the semi-dwarf characteristics–is not novel. It is a very commonplace method of breeding, which anyone familiar in the art of breeding knows. Yet the U.S. Patent and Trademark Office has issued RiceTec a broad patent, calling RiceTec's Basmati and its breeding "novel," producing a rice with "characteristics similar or superior to those of good quality Basmati rice."[13]

Patents are supposed to be granted for industrial inventions that are novel in non-obvious ways. Yet the aroma of Basmati rice, which the patent claims as new, is not novel. RiceTec's Basmati cannot be both novel and similar to traditional Basmati at the same time. The very conventional derivation of varieties through crossing is neither a novel nor a non-obvious step. In fact, the RiceTec patent treats derivation as creation and piracy as invention. The U.S. Patent Office has protected not invention but biopiracy.

RiceTec's Basmati patent illustrates the problems inherent in patenting living resources. Claiming invention for plant varieties denies both the creativity of nature on the one hand and of farmers on the other. If this false claim to invention is maintained, it could actually be used to penalize Basmati farmers for infringing on the RiceTec patent. Indian farmers who grow Basmati would be forced to pay royalties to RiceTec.

The costs to Indian agriculture would be huge. The livelihoods of 250,000 farmers growing Basmati in India and Pakistan would be jeopardized. Market monopolies would exclude the original innovators from their rightful access to local, national, and global markets.

The piracy of Basmati is just one example of how corporations are claiming "intellectual property rights" to the biodiversity and indigenous innovations of the Third World, robbing the poor of the last resources that allow them to survive outside the global marketplace. Other examples include patents on pepper, ginger, mustard, neem, and turmeric.[14]

THE THEFT OF *Kanak*

W heat is called *kanak,* or gold, in North India.

The Indian wheat economy is based on a decentralized, small-scale, local production, processing, and distribution system. Wheat and flour provide livelihoods and nutrition to millions of farmers, traders (*artis*), and processors (*Chakki Wallas,* or local flour mills). Wheat flour is also produced by millions of women working at the household level, and the rolling pin used for making flatbreads from wheat flour has always been a symbol of women's power.

The wheat economy is huge in aggregate. It generates millions of livelihoods while ensuring the availability of fresh, wholesome, sustainably produced and processed, inexpensive food. Millions of Indian farmers grow 6 billion tons of wheat every year. Most of this is sold directly to customers at the local corner store and taken to the local flour mill.

It is estimated that more than 3.5 million family-run *kirin* shops supply wheat to Indian consumers. More than 2 million small neighborhood mills produce fresh flour. While 40 million tons of wheat are traded, only 15 million tons are purchased directly as packaged flour because Indians love freshness and quality in food. Less than 1 percent of the flour consumed in India comes from packaged brands.

This decentralized, small-scale economy based on millions of producers, processors, and traders works with very little capital and very little infrastructure. People substitute for capital and infrastructure. Such a people-centered economy is, however, a block to large-scale profits for large-scale agribusiness. They are therefore eyeing the Indian wheat economy to transform it into a source of profit.

According to an industry report entitled *FAIDA* (profit), global agribusiness plans to make farmers directly dependent on them for seeds by destroying the local seed supply and by displacing the local *artis* and destroying the local flour mills. This destruction of people's access to fresh and cheap flour is described as the "modernization of the food chain." And the consumption of packaged food is described as the food culture of the rich. However, in industrialized countries, the rich eat fresh, not packaged, food. It is the poor who are forced to eat heavily processed and packaged food.

While India's wheat-and-flour economy is complex and highly developed, global agribusiness defines it as "underdeveloped" because the big players like Cargill and Archer Daniels Midland (ADM) do not control it. As the *FAIDA* report states, "The Indian wheat sector is currently at a nascent stage of development."

Agribusiness has already started trying to get Indian consumers to doubt their own quality-control systems and instead trust brand names. The corporate-controlled market they envision would generate Rs. 30 billion in revenues and Rs. 10 billion in profits, through sale of packaged brands of flour. According to industry, "the *Chakki Walla* will be a thing of the past."

The *FAIDA* report claims that 50 million jobs will be "created" by the takeover of India's local wheat economies. However, if one takes into account the 20 to 30 million farmers, 5 million *Chakki Wallas*, 5 million *artis*, 3.5 million *kirana* shops, and the households dependent on them, at least 100 million people's livelihoods and sustenance will be destroyed by the industrialization of the wheat economy.

In the United States, ADM owns 200 grain elevators, 1,900 barges, 800 trucks, and 130,000 railcars, which move wheat around without any significant employment generation by using pneumatic blowers to load and unload grain. Investment in infrastructure is used to displace people.

According to the *FAIDA* report,

> As a result of the inadequate technology used by the millers the shelf life of flour in India is typically 15 to 20 days. This is very short when compared to the six months to a year achieved in the U.S. Given the huge distances between the factory and the markets and the lengthy distribution system, the branded player has to ensure a much longer shelf life.[15]

All the positive aspects of food—freshness, local supply, low cost, low environmental impact, and high nutrition—are destroyed and replaced by negative aspects—staleness, long-distance supply, higher cost, high environmental impact, and low nutrition due to over-processing.

WTO AND THE PROMOTION
OF BIOPIRACY

*B*iopiracy is promoted by U.S. laws and World Trade Organization (WTO) agreements that globalize Western-style "intellectual property rights." There are certain distortions in U.S. law that facilitate the patenting process for companies. One such distortion is the interpretation of "prior art." It permits patents to be filed on discoveries made in the United States, whether or not identical ones already exist and are in use in other parts of the world. Unless this part of U.S. patent law (Section 102) is amended, new examples of biopiracy will continue to occur.

The General Agreement on Tariffs and Trade (GATT) on trade-related intellectual-property rights (TRIPs) calls for a system of uniform patent laws by 1999, discounting the differences in ethics and value systems of Third World nations, where life is sacred and exempt from patenting. When the TRIPs agreement was being negotiated, a Monsanto representative called it

> absolutely unprecedented in GATT. Industry has identified a major problem in international trade. It crafted a solution, reduced it to a concrete proposal, and sold it to our own and other governments.... The industries and traders of world commerce have played simultaneously the role of patients, the diagnosticians, and the prescribing physicians.[16]

Having drafted the treaty, global corporations are determined to use it. But TRIPs have been at the heart of people's resistance to the WTO. When protests and parliamentary debates resulted in the Indian government not implementing TRIPs, the U.S. government initiated a WTO dispute against India.

In 1998, the WTO ruled that India's failure to amend its patent law was illegal according to GATT. This ruling forces India to recognize U.S.-style patent regimes, and is in essence a decision against Indian democracy. India is being held guilty under the WTO "constitution," because the Indian people, the Indian Parliament, and the Indian government have acted democratically in accordance with the rights and duties bestowed on them by their national constitution.

The most effective means for challenging the RiceTec and similar patents is through the recognition and legal protection of farmers' rights. Indigenous innovation is also recognized and protected by the Convention on Biological Diversity (CBD), an international treaty signed by the world's governments at the 1992 Earth Summit in Rio, which aims to protect biodiversity, recognize countries' sovereignty over their biological wealth, and promote sustainability and equity in the use of biological resources.

The value of conserving biodiversity in general and agricultural biodiversity in particular is now undisputed. Both the CBD and the Leipzig Global Plan of Action commit governments to conserving agricultural biodiversity and recognizing farmers' rights. Governments that have agreed to the CBD are obliged to respect, preserve, maintain, and promote the wider application of knowledge, innovations, and practices of indigenous and local communities, when relevant to the conservation and sustainable use of biological diversity.

PATENTS AND POLICE STATES

*I*ntellectual-property rights and patents reorganize relationships between the human species and other species, and within the human community. Instead of the culture of the seed's reciprocity, mutuality, permanence, and exhaustless fertility, corporations are redefining the culture of the seed to be about piracy, predation, the termination of fertility, and the engineering of sterility.

The perverse intellectual-property-rights system that treats plants and seeds as corporate inventions is transforming farmers' highest duties—to save seed and exchange seed with neighbors—into crimes. Further, seed legislation forces farmers to use only "registered" varieties. Since farmers' varieties are not registered, and individual small farmers cannot afford the costs of registration, they are slowly pushed into dependence on the seed industry.

Josef Albrecht is an organic farmer in the village of Oberding in Bavaria. Not satisfied with commercially available seed, he developed his own ecological varieties of wheat. Ten other organic farmers from neighboring villages also used his wheat seeds. In 1996, the Upper Bavarian government fined Albrecht because he traded in uncertified seed. He has challenged the penalty and the Seed Act that levied it, on the grounds that the act restricts the free exercise of his occupation as an organic farmer. During the Leipzig Conference on Plant Genetic Resources, Albrecht initiated a non-cooperation movement against seed legislation, in the same Leipzig church where the democracy movement against the communist government of East Germany was organized in 1986.[17]

In Scotland, many farmers grow and sell seed potato. Until the early 1990s, they freely sold seed potato to other seed-potato growers, to merchants, and to farmers. In the 1990s, these sales became illegal. Seed-potato growers had to grow varieties under contract with the seed industry, which specified the price at which the contracting company would take back the crop, and barred growers from selling the crop to anyone. The companies started to reduce the acreage and reduce the prices. In 1994, seed potato bought from Scottish farmers for £140 was sold for more than double that price to English farmers, while the two sets of farmers were prevented from dealing directly with each other. The seed-potato growers signed a petition complaining that the stranglehold of a few companies amounted to a cartel.

The farmers also started to sell non-certified seed directly to English farmers. The seed industry claimed they were losing £4 million in seed sales through this direct trade between farmers.[18] In February 1995, the British Society for Plant Breeders sued a farmer from Aberdeenshire, who was forced to pay £30,000 in compensation to cover royalties lost to the seed industry by direct farmer-to-farmer exchange.

In the United States, direct farmer-to-farmer exchange is also illegal, as established by a case filed by the Asgrow Seed Company, now owned by Monsanto, against Dennis and Becky Winterboers. The Winterboers are farmers who own a 500-acre farm in Iowa. Since 1987, the Winterboers have derived a sizable portion of their income from selling their crops to other farmers to use as seed. In 1995, Asgrow (which has plant-variety protection for its soybean seeds) sued the Winterboers on the grounds that this direct trade violated the company's property rights. The court ruled against the Winterboers, and the Plant Variety Act, which the Winterboers had hoped would protect sales between farmers, was amended. The 1994 amendment established an absolute monopoly for the seed industry, making farmer-to-farmer exchanges and sales illegal.

Monsanto further negates farmers' rights with its "Roundup Ready Gene Agreement," the signing of which is necessary in order to purchase the company's Roundup Ready soybeans. The agreement prevents the grower from saving the seeds or from selling or supplying the seeds or material derived from them to any other person or entity. The agreement requires a payment of $5 per pound of seeds in addition to the regular price of the seeds as a "technology fee." If any clause of the agreement is violated, the grower has to pay 100 times the value of the damages. Finally, the agreement gives Monsanto the right to visit the farmer's fields, with or without the farmer's presence or permission, for three years after the agreement is signed. (As one outraged farmer commented, "We shoot intruders.")

The agreement is binding on the heirs and personal representatives of successors of growers, but growers' rights cannot be transferred without Monsanto's permission. In addition, the agreement has no liability clause. It has no reference to the performance of Roundup Ready soybeans, and Monsanto is not responsible if the seeds fail to perform as promised, or if Roundup causes ecological damage. This is especially relevant given the failure of Monsanto's genetically engineered cotton, called Bollgard, to resist damage from bollworms as advertised.

In 1998, Monsanto hired Pinkerton detectives to harass more than 1,800 farmers and seed dealers across the United States, with 475 potentially criminal "seed piracy" cases already under investigation. A group of seed-saving farmers in Kentucky, Iowa, and Illinois were forced to pay fines to Monsanto of up to $35,000 each. According to Monsanto's Scott Baucum, "We say they can pay [either of] two royalties—$6.50 at the store or $600 in court."[19]

The most dramatic case of criminalization of farmers is that of Percy Schmeiser of Saskatchewan, Canada. In a landmark case, Monsanto is suing Schmeiser for saving seeds, despite the fact that he did not buy Monsanto seeds. Rather, his fields were invaded by

Monsanto's Roundup Ready canola. Pollen from Roundup Ready crops is blowing all over the Canadian prairie and is invading farms such as Schmeiser's. But instead of paying Schmeiser for biological pollution, Monsanto is suing him for "theft" of its property.

Monsanto also sponsors a toll free "tip line" to help farmers blow the whistle on their neighbors. According to RAFI's Hope Shand, "Our rural communities are being turned into corporate police states, and farmers are being turned into criminals."[20]

1   Hope Shand, "Human Nature: Agricultural Biodiversity and Farm-Based Food Security," Rural Advancement Foundation International, 1997.
2   "Monsanto: Peddling 'Life Sciences' or 'Death Sciences'?" New Delhi: Research Foundation for Science, Technology, and Ecology (RFSTE), 1998.
3   *Wall Street Journal,* March 16, 1999.
4   Leora Broydo, "A Seedy Business," *Mother Jones Online,* www.mojones.com/news_wire.
5   Leora Broydo.
5   Leora Broydo.
6   "Monsanto: Peddling 'Life Sciences' or 'Death Sciences'?"
7   "Let Nature's Harvest Continue!" *Third World Resurgence,* No. 97.
8   Geri Guidetti, The Ark Institute, Okford, OH: 1998.
9   Monsanto's open letter on this issue is available at www.monsanto.com.
10   Emike Ohnuiki-Trerney, *Rice as Self,* Princeton: Princeton University Press, 1993, p. 9.
11   According to the CSS Haryana Agricultural University, Hissar, one of the earliest references to Basmati is made in the famous epic of *Heer . Ranjha,* written by Varis Shah in 1766.
12   Basmati Biopiracy, RFSTE, 1998.
13   U.S. Patent and Trademark Office, Patent No. 5663484.
14   Biopiracy Fact Sheets, RFSTE, 1998.
15   McKinsey and Co. and Confederation of Indian Industry, *FAIDA* report, New Delhi, India, 1999.
16   James Enyart, "A GATT Intellectual Property Code," *Les Nouvelles,* June 1990, pp. 54-56.
17   Bija Newsletter, No. 17 and 18, RFSTE, 1998.
18   Tracey Clunis Ross, "Growing Problems: The Issue of Sovereignty over Seeds," unpublished manuscript, 1995.
19   Ronnie Cummins, *Food Bytes* No. 13, October 31, 1998, p. 2.
20   "'Gene Police' Raise Farmers Fears," *Washington Post,* February 3, 1999, p. 1.

GENETIC ENGINEERING
and FOOD SECURITY

*G*enetic engineering has been sold as a green technology that will protect nature and biodiversity. However, the tools of genetic engineering are designed to steal nature's harvest by destroying biodiversity, increasing the use of herbicides and pesticides, and spreading the risk of irreversible genetic pollution.

According to the president of Monsanto, Hendrik Verfaillie, all biodiverse species that are not patented and owned by them are weeds that "steal the sunshine." Yet corporations that promote genetic engineering steal nature's harvest of diverse species, either by deliberately destroying biodiversity or by unintended biological pollution of species and ecosystems. They steal the global harvest of healthy and nutritious food. Finally, they steal knowledge from citizens by stifling independent science and denying consumers the right to know what is in their food.

"FEEDING THE WORLD"

"*F*eeding the world" is the main slogan of the biotechnology industry. In a $1.6 million European media blitz in 1998, Monsanto ran the following advertisement:

> Worrying About Starving Future Generations Won't Feed Them. Food Biotechnology Will.
>
> The world's population is growing rapidly, adding the equivalent of a China to the globe every 10 years. To feed these billion more mouths, we can try extending our farming land or squeezing greater harvests out of existing cultivation. With the planet set to double in numbers around 2030, this heavy dependency on land can only become heavier. Soil erosion and mineral depletion will exhaust the ground. Lands such as rainforests will be forced into cultivation. Fertilizer, insecticide, and herbicide use will increase globally.
>
> At Monsanto, we now believe food biotechnology is a better way forward. Our biotech seeds have naturally occurring beneficial genes inserted into their genetic structure to produce, say, insect- or pest-resistant crops.
>
> The implications for the sustainable development of food production are massive: Less chemical use in farming, saving scarce resources. More productive yields. Disease-resistant crops. While we'd never claim to have solved world hunger at a stroke, biotechnology provides one means to feed the world more effectively.
>
> Of course, we are primarily a business. We aim to make profits, acknowledging that there are other views of biotechnology than ours. That said, 20 government regulatory agencies around the world have approved crops grown from our seeds as safe.[1]

Hoechst, another self-styled "life sciences corporation," ran a similar ad in the April 16, 1999, *Financial Times*, asking us to "imagine a world where harvests grew just as fast as the population."

Ironically, Monsanto earns most of its revenue from the sale of chemicals, giving the lie to its claim that it is a "life sciences" company.[2] It attempts to cloak this fact by describing its sales of agrichemicals such as Roundup and related products as "agricultural" products rather than chemicals.

MANUFACTURING THE ILLUSION
OF SUSTAINABILITY

The "green" image that genetically engineered crops are sustainable is an illusion manufactured by corporations.

This illusion is created by several means. First, corporations attempt to portray biotechnology as an "information" technology with no material ecological impacts. As Monsanto's president has stated, "At the most basic level, then, biotechnology gives us the chance to achieve sustainability, by substituting information for stuff." What could be an easier god-trick than the argument that biotechnology achieves sustainability by "substituting information for stuff"? The material effects of genetic engineering disappear, and with them, the problem of negative ecological impacts. However, Roundup is "stuff," not information. Roundup Ready soybeans are stuff, Bollgard-cotton is stuff, the genes engineered into it are stuff, and this stuff has ecological impact.

Second, corporations promote the misinformation that transgenic crops require fewer agrichemicals. In fact, evidence shows that transgenic crops lead to increased use of hazardous chemicals (see below).

Third, when corporations describe the benefits of genetic engineering, they do so in comparison to large-scale industrial agriculture rather than to ecological, small-scale agriculture. Yet most of the world's farmers are small-scale farmers working on less than two acres, both to meet their diverse food needs and to market some of their produce.

Biotech industry consultant Clive James claims that herbicide-resistant potatoes, for instance, save farmers $6 per acre, but this is based on a farm that spends between $30 and $120 per acre on insecticide control.[3] For an organic, ecological farm, herbicide-resistant potatoes increase costs by $25 to $115 per acre, and also require increased insecticide use.

THE MYTH OF DECREASED
AGRICHEMICAL USE

*T*he development of herbicide-resistant and pest-resistant crops accounts for more than 80 percent of the biotechnology research in agriculture. However, evidence is already available that rather than controlling weeds, pests, and diseases, genetic engineering increases chemical use and can create superweeds, superpests, and superviruses.

Herbicide-resistance accounts for 71 percent of the applications of genetic engineering. Through genetically engineering herbicide resistance into crops, corporations are increasing sales of both chemicals and seeds. Monsanto's Roundup Ready soybeans are an example of such an herbicide-resistant crop.

The Roundup herbicide is Monsanto's flagship agricultural product. According to the company, Roundup, a glyphosate-based herbicide, ."destroys every weed, everywhere, economically." However, Roundup is a non-selective herbicide that does not distinguish between weeds and desirable vegetation, and thus kills all plants, which is in no way economical. Roundup effectively controls a broad range of grasses and broadleaf weeds by inhibiting EPSP synthase, an enzyme essential to a plant's growth, and establishing a road block in the plant's metabolic pathways.

According to Monsanto,

> Many of you have heard of Monsanto's Roundup herbicide. And it's very effective at killing weeds—so effective, in fact, that Roundup would control soybeans as well as weeds if it should come into contact with both.
>
> At least, that was the case until Monsanto developed Roundup Ready Soybeans. Roundup Ready Soybeans express a novel protein that allows them to thrive, even when sprayed with enough Roundup to control competing weeds.[4]

The gene inserted in Roundup Ready crops increases the amount of EPSP synthase protein in the plants, providing a detour around Roundup's roadblock. Thus, in order to prevent weeds, farmers are encouraged to grow crops they do not necessarily need or consume.

In 1995, Monsanto genetically engineered a cotton plant, named Bollgard, meant to be resistant to the common bollworm pest. This transgenic crop is meant to enable farmers to dispense with the synthetic insecticides now used to control insect pests. However, the company admits that bollworm larvae more than one quarter inch long or older than two to four days are difficult to control with Bollgard alone.[5] According to Monsanto, "if sufficient larvae of this size are present you may need to apply supplemental treatment at intervals."[6]

The company suggests maintaining a refuge for Bollgard cotton: that is, it suggests that four acres of non–Bollgard cotton crops be planted as refuge for every 100 acres of Bollgard cotton planted. In India, the small-scale farmers that dominate the cotton-growing zones would find it very difficult to maintain such refuges.

In 1997, 20 percent of the first commercial crop of Roundup Ready cotton suffered deformed bolls and bolls dropping off early. During 1998, Monsanto started field trials of Bollgard in India with the aim of marketing genetically engineered seeds by 1999-2000. A review of pesticide sprays by the farmers at various trial sites in India revealed that the use of pesticides had not stopped at all for the Bollgard crop.[7]

Experiments with some caterpillar pests of cotton have proved that some pests (for example, *Spodoptera* and *Heliothis*) can develop resistance to the toxins engineered into Bollgard. Finally, since most crops have a diversity of insect pests, insecticides may still have to be applied to transgenic crops engineered to withstand just one pest. According to an analysis by the Pesticides Trust on behalf of Greenpeace, such herbicide-resistant varieties will alter the pattern of herbicide use, but will not change the overall amounts used.[8]

THE MYTH OF INCREASED YIELDS AND RETURNS

*H*uman ingenuity has always kept harvests above population growth. As Clifford Geertz has shown by comparing 22 farming systems, biodiversity and labor intensification are the most efficient and sustainable ways of increasing yields.

As Marc Lappé and Britt Bailey report in their book *Against the Grain*, herbicide-resistant soybeans yielded 36 to 38 bushels per acre, while hand-tilled soybeans yielded 38.2 bushels per acre. According to the authors, this raises the possibility that the gene inserted into these engineered plants may selectively disadvantage their growth when herbicides are not applied. "If true, data such as these cast doubt on Monsanto's principal point that their genetic engineering is both botanically and environmentally neutral," the authors write.[9]

In any case, in the corporate-controlled food system, the same company may perform the research, sell the seeds, and provide the data about its products. Thus, the patient, diagnostician, and physician are rolled into one, and there is no objective basis of assessment of yield performance or ecological impact.

Although Monsanto's Indian advertising campaign reports a 50 percent increase in yields for its Bollgard cotton, a survey conducted by the Research Foundation for Science, Technology, and Ecology found that the yields in all trial plots were lower than what the company promised. Yields from the local, cultivated hybrid variety and Bollgard were more or less the same.

Bollgard's failure to deliver higher yields has been reported all over the world. The Mississippi Seed Arbitration Council ruled that in 1997, Monsanto's Roundup Ready cotton failed to perform as advertised, recommending payments of nearly $2 million to three cotton farmers who suffered severe crop losses.

While increased food productivity is the argument used to promote genetic engineering, when the issue of potential adverse impacts on farmers is brought up, the biotechnology industry itself argues that genetic engineering does *not* lead to increased productivity. Thus Robert Shapiro, CEO of Monsanto, while referring to Posilac (Monsanto's bovine growth hormone) in *Business Ethics,* said on the one hand that

> There is need for agricultural productivity, including dairy productivity, to double if we want to feed all the people who will be joining us, so I think this is unequivocally a good product.[10]

On the other hand, when asked about the product's economic impact on farmers, he said that it would "play a relatively small role in the process of increasing dairy productivity."

THE SOCIOECONOMIC COSTS
OF GENETICALLY ENGINEERED SEEDS

Cultivating genetically modified crops is more expensive than conventional crops because of the higher costs of the seed, technology fees, and the need for increased use of chemicals. In organic agriculture, the seeds are saved and cultivated the following season, and other necessary inputs for the seeds' cultivation are provided on the farm. When genetically engineered seeds are cultivated, all of these inputs must be paid for, and farmers will inevitably encounter serious financial troubles. Cultivating Bollgard cotton is estimated to cost Indian farmers nearly nine times more than cultivating a conventional variety. If the 21.4 million acres under cotton cultivation in India in 1997–98 were shifted to genetically engineered cotton, it would cost nearly Rs. 224.7 billion.

These increased costs can push farmers into bankruptcy and even suicide. The 1998 failure of the hybrid cotton crop in Andhra Pradesh due to pest devastation, and the subsequent suicide of farmers due to indebtedness—caused by spending nearly Rs. 12,000 per acre on pesticides—indicate how vulnerable our agricultural systems have become.

THE MYTH OF SAFE FOODS

Monsanto and other corporations repeatedly refer to their seeds and foods as having been tested for safety. But not only have no ecological or food-safety tests been conducted on genetically engineered crops and foods before commercialization; corporations have tried every means within their reach to steal the right to safe and nutritious food from citizens and consumers.

It is often claimed that there have been no adverse consequences from over 500 field releases in the United States. In 1993, for the first time, the data from the U.S. Department of Agriculture (USDA) field trials were evaluated to see whether they support these safety claims. The Union of Concerned Scientists (UCS), which conducted the evaluation, found that the data collected by the USDA on small-scale tests have little value for commercial risk-assessment. Many reports fail to even mention—much less measure—environmental risks. Of those reports that allude to environmental risk, most have only visually scanned field plots looking for stray plants or isolated test crops from relatives. The UCS concluded that the observations that "nothing happened" in those hundreds of tests do not say much. In many cases, adverse impacts are subtle and would never be registered by scanning a field. In other cases, failure to observe evidence of the risk is due to the contained conditions of the tests. Many test crops are routinely isolated from wild relatives, a situation that guarantees no out-crossing. The UCS cautioned that "care should be taken in citing the field test record as strong evidence for the safety of genetically engineered crops."[11]

All genetically engineered crops use genes that are resistant to antibiotics to help identify whether the genes that have been introduced from other organisms have been successfully inserted into the engineered crop. These marker genes can exacerbate the spread of antibiotic resistance among humans. Based on this concern, Britain rejected Ciba-Geigy's transgenic maize, which contains the weaker gene for campicillin resistance.

Many transgenic plants are engineered for resistance to viral diseases by incorporating the gene for the virus's coat protein. These viral genes may cause new diseases. New broad-range recombinant viruses could arise, causing major epidemics.

Upon consumption, the genetically engineered DNA of these foods can break down and enter the blood stream. It has long been assumed that the human gut is full of enzymes that can rapidly digest DNA. But in a study designed to test the survival of viral DNA in the gut, mice were fed DNA from a bacterial virus, and large fragments were found to survive passage through the gut and to enter the bloodstream.[12] Further studies indicate that the ingested DNA can end up in the spleen and liver cells as well as in white blood cells.[13]

Within the gut, vectors carrying antibiotic-resistance markers may also be taken up by the gut bacteria, which would then serve as a mobile reservoir of antibiotic-resistance genes for pathogenic bacteria. Horizontal gene transfer between gut bacteria has already been demonstrated in mice and chickens and in human beings.[14]

When L-tryptophan, a nutritional supplement, was genetically engineered and first marketed, 37 people died and 1,500 people were severely affected by a painful and debilitating circulatory disorder called eosinophilia myalgia.[15] When a gene from the Brazil nut was inserted into soybeans to increase their protein levels, the transgenic soybeans also contained the nut's allergenic properties.[16]

Greenpeace and other non-governmental organizations have revealed that soybean plants sprayed with Roundup are more estrogenic and could act as hormone or endocrine-system disrupters. Dairy cows that consume Roundup Ready soybeans produce milk with higher fat levels than cows that eat regular soybeans.

THE MYTH OF FOOD SECURITY

*T*he Green Revolution narrowed the basis of food security by displacing diverse nutritious food grains and spreading monocultures of rice, wheat, and maize. However, the Green Revolution focused on staple foods and their yields. The genetic engineering revolution is undoing the narrow gains of the Green Revolution both by neglecting the diversity of staples and by focusing on herbicide resistance, not higher yields.

According to Clive James, transgenic crops are not engineered for higher yields. Fifty-four percent of the increase in transgenic crops is for those engineered for herbicide resistance, or, rather, the increased use of herbicides, not increased food. As an industry briefing paper states, "The herbicide tolerant gene has no effect on yield per se."[17] Worldwide, 40 percent of the land under cultivation by genetically engineered crops is under soybean cultivation, 25 percent under corn,

13 percent under tobacco, 11 percent under cotton, 10 percent under canola, and 1 percent each under tomato and potato. Tobacco and cotton are non-food commercial crops, and crops such as soybeans have not been food staples for most cultures outside East Asia. Such crops will not feed the hungry. Soybeans will not provide food security for *dal*-eating Indians, and corn will not provide security in the sorghum belt of Africa.

The trend toward the cultivation of genetically engineered crops indicates a clear narrowing of the genetic basis of our food supply. Currently, there are only two commercialized staple-food crops. In place of hundreds of legumes and beans eaten around the world, there is soybean. In place of diverse varieties of millets, wheats, and rices, there is only corn. In place of the diversity of oil seeds, there is only canola.

These crops are based on expanding monocultures of the same variety engineered for a single function. In 1996, 1.9 million acres around the world were planted with only two varieties of transgenic cotton, and 1.3 million acres were planted with Roundup Ready soybeans. As the biotechnology industry globalizes, these monoculture tendencies will increase, thus further displacing agricultural biodiversity and creating ecological vulnerability.

Further, by forcing the expansion of non-food crops such as tobacco and cotton, transgenic crops result in fewer acres in food production, aggravating food insecurity.

THE DESTRUCTION OF BIODIVERSITY

*I*n Indian agriculture, women use up to 150 different species of plants (which the biotech industry would call weeds) as medicine, food, or fodder. For the poorest, this biodiversity is the most important resource for survival. In West Bengal, 124 "weed" species collected from rice fields have economic importance for local farmers. In a Tanzanian village, over 80 percent of the vegetable dishes are prepared from uncultivated plants.[18] Herbicides such as Roundup and the transgenic crops engineered to withstand them therefore destroy the economies of the poorest, especially women. What is a weed for Monsanto is a medicinal plant or food for rural people.

Since biodiversity and polycultures are an important source of food for the rural poor, and since polycultures are the most effective means of soil conservation, water conservation, and ecological pest and weed control, the Roundup Ready technologies are in fact a direct assault on food security and ecological security.

THE RISKS OF GENETIC POLLUTION

*G*enetically engineered crops increase chemical use and add new risks of genetic pollution. Herbicide-resistant crops are designed for intensive use of herbicides in agriculture. But they also create the risks of weeds being transformed into "superweeds" by the transfer of herbicide-resistant traits from the genetically engineered crops to closely related plants.

Research in Denmark has shown that oilseed rape genetically engineered to be herbicide-tolerant could transmit its introduced gene to a weedy natural relative through hybridization. Weedy relatives of rape are now common in Denmark and throughout the world. Converting these "weeds" into "superweeds" that carry the gene for herbicide-resistance would provoke high crop losses and increasing use of herbicides. For these reasons, the European Union has imposed a *de facto* moratorium on the commercial planting of genetically engineered crops.

In many cases, the weeds that plague cultivated crops are relatives of the crops themselves. Wild beets have been a major problem in European sugar-beet cultivation since the 1970s. Given the gene exchange between weedy beets and cultivated beets, herbicide-resistant sugar beets could only be a temporary solution. [19]

Superweeds could lead to "bioinvasions," displacing local diversity and taking over entire ecosystems. The problem of invasive species is being increasingly recognized as a major threat to biodiversity. Monsanto's claim that products such as Roundup Ready soybeans will reduce herbicide use is false because it does not take into account the introduction of such engineered plants in regions where herbicides are not used in agriculture and where native diversity of soybeans exists. China, Taiwan, Japan, and Korea are regions where soybeans have evolved and where wild relatives of cultivated soybeans are found. In these regions, Monsanto's Roundup Ready soybeans would increase herbicide use and "pollute" the native biodiversity by transferring herbicide-resistant genes to wild plants. This could lead to new weed problems and loss of biodiversity. Moreover, since the Third World is the home to most of the world's biodiversity, the risks of genetic pollution in Third World countries are even more profound.

Herbicide-resistant transgenic crops can also become weeds when seeds from those crops germinate after harvest. More herbicides will have to be applied to eliminate these "volunteer plants."

TOXIC PLANTS: A RECIPE FOR SUPERPESTS

*T*he bacterium *Bacillus thuringiensis* (Bt) was isolated from soil in 1911. Since 1930, it has been available as an organic form of pest control. Organic farmers have stepped up its use since the 1980s.

Monsanto and other "life sciences" corporations developed a technique of inserting the toxin-producing gene from the Bt bacteria into plants. This particular Bt gene produces a toxin that disables insects, and the genetically engineered Bt plants are thus able to produce their own pesticide. Genetically engineered Bt-crops have been cultivated commercially since 1996.

While Monsanto sells Bt-crops with the claim that they will reduce pesticide use, Bt-crops can actually create "superpests" and increase the need for pesticides. Bt-crops continuously express the Bt toxin throughout their growing season. Long-term exposure to the toxins promotes the development of resistance in insect populations. This kind of exposure could lead to selection for resistance in all stages of the insect pest on all parts of the plant for the entire season.

Due to these risks of encouraging pest resistance, the U.S. Environmental Protection Agency (EPA) offers only conditional and temporary registration for Bt-crops. The EPA requires a 4 percent refuge for Bt cotton—i.e., 4 percent of the cotton in a Bt-cotton field must be conventional and not express the Bt toxin. The conventional cotton acts as a refuge for insects to survive and breed in order to keep the overall level of resistance in the population low.

While the Monsanto propaganda states that farmers will not have to use pesticides, the reality is that the management of resistance requires continued use of non-Bt cotton and pesticide sprays. And even with a 4 percent refuge, insect resistance will evolve in as few as three to four years. Already eight species of insects have developed resistance to Bt toxins, including diamond black moth, Indian meal moth, tobacco budworm, colorado potato beetle, and two species of mosquitoes.[20]

Even if Bt-crops do repel some pests, most crops have a diversity of insect pests. Insecticides will still have to be applied to control pests that are not susceptible to Bt's toxin. Beneficial species such as birds, bees, butterflies, and beetles, which are necessary for pollination and which through the prey-predator balance also control pests, may be threatened by Bt-crops.[21] Soil-inhabiting organisms that degrade the toxin-contaminated organic matter can be harmed by the toxin. Nothing is known of the impact on human health when Bt-crops such as potato and corn are eaten, or on animal health when oilcake from Bt-cotton or fodder from Bt-corn is consumed as cattle feed.

THE POLITICS OF BIOSAFETY

*B*iosafety, or the prevention of biohazards caused by genetic engineering, is emerging as the most important environmental and scientific issue of our time. Biosafety issues are intimately linked to the politics of science, and to the conflicting perspectives of different scientific cultures and traditions.

One conflict is between the ecological sciences that assess the impact of genetic engineering on the environment and on human health, and reductionist sciences that promote production based on genetic engineering.

A second conflict is between private-interest and public-interest science. When the techniques of recombinant DNA were emerging in the late 1970s and 1980s, the crippled organisms that resulted from the experiments were not meant to survive in the environment. The main practitioners during this phase were university scientists, and they themselves called for a moratorium on recombinant DNA research.

During the 1980s and 1990s, scientists who had developed genetic engineering techniques left universities to start biotechnology firms. During this phase, concerns for safety were sidelined by the promise of biotech miracles. Today, genetically engineered organisms are being released for production and consumption on global markets, and small, start-up biotech firms are being bought up by giant chemical corporations.

The biosafety issues that were outlined by university scientists using crippled organisms are very different from those posed by robust organisms being produced by transnational corporations for global markets. These issues interfere in the market expansion of genetic engineering in agriculture, and thus industry has attempted to suppress the debate in four main ways.

First, they invoke a call to "sound science," which they equate with industry-friendly science, and treat industry-independent science as "junk science." "Sound science" has become like a mantra for banishing safety regulations. This was the phrase used by the industry in a letter to President Clinton at the G7 Summit in Denver in 1997.[22] It is the language of *The Wall Street Journal* editorial accusing Europe of practicing "junk science" by banning the import of hormone-fed beef, and referring to the World Trade Organization (WTO) decision against the ban as "real science."[23] According to the U.S. agricultural secretary, Dan Glickman, who has stated categorically that the United States will stand behind its genetically engineered foods and oppose any European labeling requirements as a trade violation,

> We've got to make sure that sound science prevails, not what I call historic culture, which is not based on sound science. Europe has a much greater sensitivity to the culture of food as opposed to the science of food. But in the modern world, we just have to keep the

pressure on the science. Good science must prevail in these decisions.[24]

However, the conflict over genetically engineered crops and foods is not a conflict between "culture" and "science." It is between two cultures of science: one based on transparency, public accountability, and responsibility toward the environment and people, and another based on profits and the lack of transparency, accountability, and responsibility.

Second, the industry claims that there is "substantial equivalence" between genetically engineered products and natural ones. When corporations claim monopoly rights to seeds and crops, they refer to genetically modified organisms (GMOs) as "novel." When the same corporations want to disown risks by stifling safety assessment and analysis of hazards, they refer to transgenic organisms as being substantially equivalent to their naturally occurring counterparts. The same organism cannot be both "novel" and "not novel." This ontological schizophrenia is a convenient construct to create a regime of absolute rights and absolute irresponsibility. Through the WTO, the ontological schizophrenia is being spread from the United States to the rest of the world.

The genetic engineering guidelines of the Food and Drug Administration (FDA) are based on the assumption that GMOs behave like their naturally occurring counterparts. The guidelines are also based on the assumption that "genetically engineered organisms have greater predictability compared to species evolved by traditional techniques." Neither of these assumptions is true. GMOs do not behave like their naturally occurring counterparts, and the behavior of GMOs is highly unpredictable and unstable.

For example, naturally occurring *Klepsiella planticola* does not kill plants, but, as research at the University of Oregon has shown, the genetically engineered *Klepsiella* was lethal to crops.[25] The naturally occurring *Bacillus thuringiensis* has not contributed to the evolution of resistance in pests, but the genetically engineered Bt-crops create rapid resistance evolution because the Bt toxin is expressed in *every* cell of the plant, *all* the time. Thus the assumption of "substantial equivalence" does not hold.

The assumption of "predictability" is also totally false. While genetic engineering makes the *identification* of the gene to be transferred into another organism more predictable, the ecological *behavior* of the transferred gene in the host genome is totally unpredictable. A transgenic yeast, which was engineered to ferment faster, accumulated a certain metabolite at toxic levels. Between 64 and 92 percent of the first generation of transgenic tobacco plants is unstable. Petunias do not have unstable coloring, but genetically engineered petunias change their color unpredictably due to "gene silencing."[26]

In 1998, when Dr. Arpad Pusztai concluded from experiments on rats that there was a lack of equivalence in both composition and metabolic consequences between genetically engineered and conventional potatoes, he was sacrificed to protect corporate control and profits. Pusztai was suspended by his lab, accused of scientific fraud, and banned from speaking to the media about his results. In 1999, 20 scientists from 14 countries examined the Pusztai report and accused his employer, the Rowett Institute in Scotland, of bowing to public pressure. Claims of a cover-up were reinforced when it was revealed that Rowett had received £140,000 of funding from Monsanto. In 1999, Dr. S.W.B. Even, a senior pathologist at the University of Aberdeen, provided conclusive evidence supporting Pusztai's findings.[27]

Third, as has been discussed above, the biotech industry further attempts to elide biosafety issues by describing contained, artificially constructed experiments as "field trials" that prove safety, and by arguing that the labeling of genetically engineered foods, guaranteeing consumers the "right to know" and the "right to choose," interferes with free trade.

Fourth and finally, the ultimate step in total control over the food system is the attempt by the USDA to destroy the organic option for farmers and consumers. If adopted and implemented, the USDA policy would outlaw genuine organic production all over the world.

Under this policy, the USDA will allow fruit and vegetables that have been genetically engineered, irradiated, treated with additives, and raised on contaminated sewage sludge to be labeled "organic."

"Organic" livestock can be housed in batteries, fed with the offal of other animals, and injected with biotics.

Further, the policy prohibits the setting of any standards higher than those established by the department. Farmers will, in other words, be forbidden by law from producing and selling good, safe food. As Thames University professor George Monbiot writes, "Organic produce, in the brave new world of American oligopoly, will be virtually undistinguishable from conventionally toxic food."[28] To date, the policy has been stalled by virtue of a major citizen mobilization against it.

THE SUBVERSION OF BIOSAFETY LAWS

*T*he United Nations Convention on Biological Diversity (CBD) outlined international biosafety laws. A small team from the Third World Network worked closely with Third World governments to introduce these rules into the CBD. Article 19.3 of the Convention states,

> the Parties shall consider the need for ... appropriate procedures, including, in particular, advance informed agreement, in the field of the safe transfer, handling, and use of any living modified organism resulting from biotechnology that may have adverse effect on the conservation and sustainable use of biological diversity.

The language of "living modified organism" was introduced by the United States in place of "genetically modified organism" to neutralize public concern about genetic engineering. "Living modified organism" applies to all products of conventional breeding, not just genetically engineered species. Then-President George Bush refused to sign the CBD because, according to him, it would interfere with the growth of the $50 billion U.S. biotech industry.

In spite of not being a party to the CBD, the United States has been present at every negotiation regarding the convention. It tried to undo to work of Panel IV, set up by the United Nations to implement CBD articles on biosafety. Although environmentalists succeeded in keeping the issue of biosafety alive for seven years despite U.S. intransigence and irrationality, a small group of countries including the United

States killed the Biosafety Protocol in 1999, on the grounds that it would interfere with WTO free-trade rules.

CULTIVATING DIVERSITY

*I*n the mountain farming systems of the Garhwal Himalaya, there is a particular cropping pattern called *baranaja,* which means literally "12 seeds." The seeds of 12 or more different crops are mixed and then randomly sown in a field fertilized with cow dung and farmyard manure. Care is taken to balance the distribution of the crops in each area of the field. After sowing, the farmer transplants crops from one area of the field to another area in order to maintain an even distribution of the crops. As in other cultivation practices, constant weeding is necessary. The crops are all sown in May, but are harvested at different times, from late August to early November, thus ensuring a continuous food supply for the farmer during this period and beyond. The different crops have been selected by the farmers over the ages by observing certain relationships between plants, and between plants and soil. For example, the *rajma* creeper will climb only on the *marsha* plant and on no other plant in the field.

The symbiotic relationships between different plants contribute to the increased productivity of the crops. When farmers cultivate *baranaja,* they get higher yields, diverse outputs, and a better market price for their produce than when they cultivate a monoculture of soybeans. Soybeans sell for only Rs. 5 per kilogram, whereas *jakhia,* one of the *baranaja* crops that matures the earliest, sells for Rs. 60 per kilogram.

Cultivating diversity can therefore be part of a farming strategy for high yields and high incomes. But since these yields and incomes are from diverse crops, centralized commercial interests are not interested in them. For them, uniformity and monocultures are an imperative. However, from the point of view of small farmers, diversity is both highly productive and sustainable.[29]

GENETIC ENGINEERING
AND FOOD SECURITY

*D*iversity and high productivity go hand in hand if diverse outputs are taken into account and the costs of external inputs are added to the cost of inputs. The monoculture paradigm focuses on yields of single commodities and externalizes the costs of chemicals and energy. Inefficient and wasteful industrial agriculture are hence presented as efficient and productive.[30]

The myth of increasing yields is the most common justification for introducing genetically engineered crops in agriculture. However, genetic engineering is actually leading to a "yield drag." On the basis of 8,200 university-based soybean trials in 1998, it was found that the top Roundup Ready soybean varieties had 4.6 bushels per acre, or yields 6.7 percent lower than the top conventional varieties. As environmental consultant Dr. Charles Benbrook states,

> In 1999, the Roundup Ready Soybean yield drag could result in perhaps a 2.0 to 2.5 percent reduction in national average soybean yields, compared to what they would have been if seed companies had not dramatically shifted breeding priorities to focus on herbicide tolerance. If not reversed by future breeding enhancements, this downward shift in soybean yield potential could emerge as the most significant decline in a major crop ever associated with a single genetic modification.[31]

Research on trials with Bt cotton in India also showed a dramatic reduction in yields: in some cases as high as 75 percent.[32]

As criticism of biotechnology's emphasis on herbicide-resistant crops and crops that produce toxins grows, the biotechnology industry has started to talk of engineering crops for nitrogen fixing, salinity tolerance, and high nutrition instead. However, all these traits already exist in farmers' varieties and farmers' fields. Legumes and pulses intercropped with cereals fix nitrogen. In coastal ecosystems, farmers have evolved a variety of salt-tolerant crops. We do not need genetic engineering to give us crops rich in nutrition. Amaranth has nine times more calcium than wheat and 40 times more calcium than rice.

Its iron content is four times higher than that of rice, and it has twice as much protein. *Ragi* (finger millet) provides 35 times more calcium than rice, twice as much iron, and five times more minerals. Barnyard millet contains nine times more minerals than rice. Nutritious and resource-prudent crops such as millets and legumes are the best path of food security.

Biodiversity already holds the answers to many of the problems for which genetic engineering is being offered as a solution. Shifting from the monoculture mind to biodiversity, from the engineering paradigm to an ecological one, can help us conserve biodiversity, meet our needs for food and nutrition, and avoid the risks of genetic pollution.

1   "Monsanto: Peddling 'Life Sciences' or 'Death Sciences'?" New Delhi: Research Foundation for Science, Technology, and Ecology (RFSTE), 1998.

2   "Monsanto: Peddling 'Life Sciences' or 'Death Sciences'?" p. 12.

3   Clive James, "Global Status of Transgenic Crops in 1997," *ISAAA Briefs,* 1997, p. 20.

4   International Association of Plant Breeders, "Feeding the 8 billion and Preserving the Planet," NYON, Switzerland.

5   Monsanto promotional material, 1996.

6   Monsanto, Bollgard, 1996.

7   Vandana Shiva, Afsar Jafri, and Ashok Emani, "Globalization of the Seed Sector," Bombay: EPW, 1999.

8   International Agricultural Development, 1998.

9   Marc Lappé and Britt Bailey, *Against the Grain: Biotechnology and the Corporate Takeover of Your Food,* Monroe, ME: Common Courage Press, 1998.

10   Interview with Robert Shapiro, *Business Ethics,* January-February 1996, p. 47.

11   Margaret Mellon and Jane Rissler, *Risks of Genetically Engineered Crops,* Cambridge, MA: MIT Press, 1996.

12   Mae Wan Ho, *Genetic Engineering: Dream or Nightmare,* Bath, U.K.: Gateway Books, 1998, p. 165.

13   Philip Cohen, "Can DNA in food finds its way into cells?" *New Scientist,* January 4, 1997, p. 14.

14   Mae Wan Ho.

15   Lappé and Bailey, p. 134.

16   J. A. Nordlee et al., "Identification of a Brazil Nut Allergen in Transgenic Soybeans," *The New England Journal of Medicine,* No. 334, 1996, pp. 688–92.

17   Clive James, p. 14.

18   Jane Rissler and Margaret Mellon, *The Ecological Risks of Engineered Crops,* Cambridge, MA: MIT Press, 1996.

19   P. Bondry, M. Morchen, et al., "The origin and evolution of weed beets: consequences for the breeding and release of herbicide resistant transgenic sugar beets," *Theoretical and Applied Genetics,* No. 87, 1993, pp. 471–78.

20   Miguel Altieri, "Ecological Impact of Genetic Engineering," unpublished paper, 1998.

21   Vandana Shiva and Afsar H. Jafri, "Seeds of Suicide," RFSTE, 1998.

22   Letter of U.S. Agribusiness to President Clinton at G7 Summit, Denver, June 18, 1997.

23   *Wall Street Journal* Editorial, November 6, 1997.

24   Dan Glickman, quoted in Vandana Shiva, *Betting on Biodiversity,* New Delhi: RFSTE, 1998, p. 45.
25   Report of the Independent Group of Scientific and Legal Experts on Biosafety, 1996.
26   Report of the Independent Group, 1996.
27   COST 98 Action (European Union Program) in Lund, Sweden, November 25–27, 1998.
28   George Monbiot, "Food Fascism," *Guardian,* March 3, 1998.
29   Research Foundation for Science, Technology, and Natural Resource Policy, "Cultivating Diversity: Biodiversity Conservation and the Politics of the Seed," New Delhi, 1993.
30   Vandana Shiva, "Biodiversity-Based Productivity," New Delhi: RFSTE, 1998; and Peter Rosset and Miguel Altieri, "The Multiple Functions and Benefits of Small Farm Agriculture," International Forum on Agriculture, San Francisco, 1999.
31   Charles Benbrook, "Evidence of the Magnitude and Consequences of the Roundup Ready Soybean Yield Drag from University-Based Varietal Trials in 1998," InfoNet Technical Paper, No. 1, Sandpoint, Ohio: July 13, 1999, p. 1.
32   Vandana Shiva et al., "Globalization and Seed Security: Transgenic Cotton Trials," *EPW,* Vol. 34, No. 10–11, March 6–19, 1999, p. 605.

RECLAIMING
FOOD *Democracy*

*F*ood democracy is an imperative in this age of food dictatorship, in which a handful of global corporations control the global food supply and are reshaping it to maximize their profits and their power. Food democracy is being created through a new solidarity between environmental democracy and sustainable-agriculture movements, farmers' movements, consumer movements, and new movements of public-interest scientists.

The central concern of citizens' movements, North and South, is creating democratic control over the food system to ensure sustainable and safe production and equitable distribution and access to food. Democratic control over food requires the reining in of the unaccountable power of corporations. It involves replacing the "free trade" order of corporate totalitarianism with an ecological and just system of food production and distribution, in which the earth is protected, farmers are protected, and consumers are protected.

Industrial agriculture in general and genetic engineering in agriculture in particular increase commodity production for the market by taking away nature's share of nutrition, and by increasing external inputs such as pesticides, herbicides, and synthetic fertilizers. Returning to nature and her species their share of nutrition is not just an ethical and ecological imperative; it is essential for maintaining food productivity for humans.

Industrial agriculture based on a reductionist, fragmented, and competitive worldview interprets partnerships, cooperation, and mutual help as competition. Instead of viewing cows and earthworms as our helpers in food production, it views them as making competing demands on food, and thus views the denial of their right to nutrition as a gain in human nutrition. Thus, in breeding, the yield of grain is increased at the cost of straw. Food for humans is increased at the cost of food for cows and earthworms.

Reclaiming democracy in food production implies reclaiming the rights of all species to their share of nutrition and, through this ecological step, reclaiming the right of all people to food rights, including future generations. A food democracy that is inclusive is the highest form of equity and democracy. Such a democracy can feed us abundantly because other species do not feed themselves at our cost; they feed us while they feed themselves.

MOVEMENTS FOR
ORGANIC AGRICULTURE

*I*n India, the poorest peasants are organic farmers because they could never afford chemicals. Today, they are joined by a growing international organic movement that consciously avoids chemicals and genetic engineering. A U.S. nationwide survey released in November 1998 by the agribusiness-affiliated International Foods Safety Council found that 89 percent of U.S. consumers think food safety is a "very important" national issue—more important than crime prevention. Seventy-seven percent were changing their eating habits due to food-safety concerns.[1] A *Time* magazine poll published in its January 13, 1999, issue found that 81 percent of U.S. consumers believe genetically engineered food should be labeled. Fifty-eight percent of consumers said they would not eat genetically engineered foods if they were labeled. In 1998, over 5 billion dollars worth of organic food was consumed in the United States, where the organic market is growing at 25 percent annually.

In India, ARISE, the national network for organic agriculture, holds village-level courses throughout the country to support farmers wanting to give up chemical addiction. Ecological and organic agriculture is often referred to in India as *ahimsic krishi,* or "non-violent agriculture," because it is based on compassion for all species and hence the protection of biodiversity in agriculture.

While organic agriculture is a low-input, low-cost option, and hence an option for the poor, it is often presented as a "luxury of the rich." This is not true. The cheapness of industrially produced food and expensiveness of organic foods does not reflect their cost of production but the heavy subsidies given to industrial agriculture. The International Federation of Organic Agriculture Movements has been working toward the global democratization of organic agriculture.

MOVEMENTS AGAINST GENETIC ENGINEERING

*I*n November 1998, farmers in Andhra Pradesh and Karnataka in India uprooted and burnt Monsanto's Bollgard crops planted in trial fields. In February 1998, a suit calling for an end to genetic engineering trials and a ban on genetically engineered food imports, filed by environmentalists and farmers, was admitted to the Supreme Court in India.

In Britain, a movement called Genetix Snowball, launched in 1998 when five women uprooted Monsanto's crops in Oxfordshire, removes genetically engineered crops from trial sites to protect the environment. In February 1999, an alliance of U.K. farm groups, consumer groups, development groups, and environmental groups launched a campaign for a "Five-Year Freeze" on genetic engineering.

In 1993 in Switzerland, a grassroots-funded organization called the Swiss Working Group on Genetic Engineering collected 111,000 names in favor of a referendum to ban genetic engineering. The biotech industry hired a public relations company for $24 million to defeat the referendum, which was outvoted by a margin of two to one in June 1998. But the debate is far from over. A similar referendum was organized by Greenpeace and Global 2000 in Austria.

In Germany, resistance to genetic engineering is led by the Gen-Ethisches Network, BUND, and a grassroots initiative called Food from the Genetics Laboratory.

In Ireland, the Gaelic Earth Liberation Front dug up a field of Roundup Ready sugar beet at Ireland's Teagase Research Centre at Oakport. In France, farmers of Confédération Paysanne destroyed Novartis's genetically engineered seeds. Subsequently, France imposed a two-year moratorium on transgenic crops.

Throughout Europe, bans and moratoriums on genetic engineering, in response to growing citizen pressure, are increasing. In July 1998, citizens from across the world met in St. Louis, Missouri, where Monsanto's headquarters are located, for a conference on "biodevastation" and to conduct protests at Monsanto. This gathering launched a new global movement of citizens against global corporations trying to control the very basis of our lives.

SAVE THE SEED

*A*nother attempt to reclaim food democracy has been through reclaiming the seed from the destructive control of corporations. For more than a decade, Indian environmentalists and farmers have built Navdanya, the movement for saving seed.

In periods of injustice and external domination, when people are denied economic and political freedom, reclaiming freedom requires peaceful non-cooperation with unjust laws and regimes. This peaceful non-cooperation with injustice has been the democratic tradition of India and was revived by Mohandas Gandhi as *satyagraha*. Literally, *satyagraha* means the struggle for truth. According to Gandhi, no tyranny can enslave people who consider it immoral to obey laws that are unjust. As he stated in *Hind Swaraj,* "As long as the superstition that people should obey unjust laws exists, so long will slavery exist. And a non-violent resister alone can remove such a superstition."

On March 5, 1998, the anniversary of Gandhi's call for the salt *satyagraha,* a coalition of more than 2,000 groups started the *bija satyagraha,* a non-cooperation movement against patents on seeds and plants.

Seed is a vital resource for the survival of life anywhere. Seed is a unique and priceless gift of nature evolved, bred, and used by farmers over millennia to produce food for the people. Farmers select and save the best seeds from a good crop to plant them again the next season. This seed-selection, -saving, and -replanting cycle has continued since the beginning of agriculture.

The salt *satyagraha* embodied India's refusal to cooperate with the unjust salt laws and was an expression of India's quest for freedom with equity. The *bija satyagraha* is our refusal to accept the colonization of life through patents and perverse technologies, and the destruction of the food security by the free trade rules of the World Trade Organization. It is an expression of the quest for freedom for all people and all species, and an assertion of our food rights.

Navdanya's aim is to cover the country with seed banks and organic farming initiatives. Navdanya will not recognize patents on life, including patents on seed. It aims to build a food and agriculture system that is patent-free, chemical-free, and free of genetic engineering. This movement will reclaim our food freedom by strengthening our partnership with biodiversity.

THE MONSANTO CAMPAIGN

*B*ecause of the nationwide awareness of genetic engineering and Monsanto created by the "Monsanto, Quit India" movement, in 1999, news of Monsanto's genetic-engineering trials in India was leaked to the press. These trials were being carried out in 40 locations in nine states. Since agricultural decisions are supposed to be made by regional governments, state agricultural ministers objected that they had not been consulted on the trials. They released the locations of the trial sites, and immediately farmers in Karnataka and Andhra Pradesh up-rooted and burned genetically engineered crops.

In Andhra Pradesh, the farmers also got a resolution passed through the regional parliament and put pressure on the government to ban the trials. After the first uprooting by farmers, the government itself up-rooted the Bt-crop in other locations.

BUILDING ALLIANCES

*T*he global movement for food democracy is building broad-based alliances—alliances between public-interest scientists and the people, between producers and consumers, between North and South. Solidarity and synergy between diverse groups is neces-sary because the corporate push for genetic engineering raises is-sues of democracy at many levels.

Public scientists who have worked on the science of ecological im-pact have been an important part of this movement. In 1994, Brian Goodwin, the eminent development biologist; Tewolde Egziabher, Ethiopia's environment secretary; Nicanor Perlas of the Philippines; and I proposed a meeting of scientists working on non-reductionist ap-proaches to biology. The Third World Network in Penang generously offered to host the meeting. The team of public scientists who gathered in Penang—Mae Wan Ho, Christine Von Weiszacker, Beatrix Tappeser, Peter Wills, and Jose Lutzenberger, along with Elaine Ingham, Beth Burrows, Terje Traavik, and others—has played a key role in raising ecological and safety issues.

Without these scientists' solidarity with citizens' movements, in-dustry's attempt to polarize the debate as if it were between "informed scientists" and "uninformed citizens," or between "reason and emo-tion," would have been successful. The protests would have been brushed aside, and commercialization of genetically engineered organisms would have continued without any question or pause.

Solidarity between producers and consumers is also necessary. Since most people in the South are farmers, and only 2 percent of the world's farmers survive in the North, movements for food democracy will take the shape of consumer movements in the North and both farmers' and consumer movements in the South.

Our movements for the recovery of the biodiversity and intellectual commons are the basis of the democratization of the food system. On the one hand, refusal to recognize life's diversity as corporate inventions and hence corporate property is a positive recognition of the intrinsic value of all species and their self-organizing capacity. On the other hand, the refusal to allow privatization of living resources through patents is a defense of the right to survival of the two-thirds majority that depends on nature's capital and is excluded from markets because of its poverty. It is also a defense of cultural diversity, since the majority of diverse cultures do not see other species and plants as "property" but as kin. This larger democracy of life, based on the earth democracy, or what we call *vasudhaiva kutumbakum,* is the real force of resistance against the brute power of the "life sciences industry," which is pushing millions of species to extinction and millions of people to the edge of survival.

If we can still imagine food freedom and work to make it real in our everyday lives, we will have challenged food dictatorship. We will have reclaimed food democracy.

1   Ronnie Cummins, *Food Bytes,* No. 16, January 28, 1999.

AFTERWORD

$\mathcal{T}$he failure of the World Trade Organization (WTO) Third Ministerial meeting in Seattle in late 1999 was a historic watershed. The rebellion on the streets and the rebellion within the WTO negotiations launched a new democracy movement, with citizens from across the world and the governments of the South refusing to be bullied and excluded from decisions in which they have a rightful share.

In Seattle, fifty thousand citizens from all walks of life and all parts of the world protested peacefully on the streets for four days to ensure that there would be no new round of trade negotiations for accelerating and expanding the process of globalization.

Trade ministers from Asia, Africa, Latin America, and the Caribbean refused to join hands to provide support to a "contrived" consensus since they had been excluded from the negotiations being undertaken in the "green room" process behind closed doors. As long as the conditions of transparency, openness, and participation were not ensured, developing countries would not be party to a consensus. Their refusal will make it difficult for industrialized countries to bulldoze decisions in future trade negotiations.

Seattle had been chosen by the United States to host the Third Ministerial conference because it is the home of Boeing and Microsoft, and symbolizes the corporate power that WTO rules are designed to protect and expand. Yet the corporations stayed in the background, and proponents of free trade and the WTO were forced to go out of their way to say that WTO was a "member-driven" institution controlled by governments who made democratic decisions.

But the WTO has earned itself names such as the World Tyranny Organization because it enforces tyrannical, anti-people, anti-nature decisions to enable corporations to steal the world's harvests through secretive, undemocratic structures and processes. The WTO institutionalizes forced trade, not free trade, and, beyond a point, coercion and the rule of force cannot continue.

The WTO tyranny was apparent in Seattle both on the streets and inside the Washington State Convention Center, where the negotiations were taking place. Intolerance of democratic dissent, a hallmark of dictatorship, was unleashed in full force. While the trees and stores were lit up for Christmas festivity, the streets were barricaded and blocked by the police, turning the city into a war zone. Non-violent protestors, including young people and old women, labor activists and environmental activists, and even local residents, were brutally beaten, sprayed with tear gas, and arrested by the hundreds.

The media has referred to the protestors as "power mongers" and "special interest" groups. Globalizers, such as Scott Miller of the U.S. Alliance for Trade Expansion, said that the protestors were acting out of fear and ignorance.

But the thousands of youth, farmers, workers, and environmentalists who marched the streets of Seattle in peace and solidarity were not acting out of ignorance and fear; they were outraged because they know how undemocratic the WTO is, how destructive its social and ecological impacts are, and how the rules of the WTO are driven by the objectives of establishing corporate control over every dimension of our lives—our food, our health, our environment, our work, and our future.

When labor joins hands with environmentalists, when farmers from the North and farmers from the South make a common commitment to say "no" to genetically engineered crops, they are not acting in their special interests. They are defending the common interests and common rights of all people, everywhere. The divide-and-rule policy, which has attempted to pit consumers against farmers, the North against the South, labor against environmentalists, has failed.

RECLAIMING THE STOLEN HARVEST

*C*itizens went to Seattle with the slogan "No new round, turn-around." They were successful in blocking a new round. The next challenge is to turn the rules of globalization and free trade around, and make trade subservient to the higher values of the protection of the earth and people's livelihoods.

As this book illustrates, against all odds, millions of people from across the world have been putting the principles of ecological agriculture into practice. The post-Seattle challenge is to change the global trade rules and national food and agricultural policies so that these practices can be nurtured and spread, and so that ecological agriculture, which protects small farms and peasant livelihoods, and produces safe food, is not marginalized and criminalized. The time has come to reclaim the stolen harvest and celebrate the growing and giving of good food as the highest gift and the most revolutionary act.

—Vandana Shiva
New Delhi, India
December 1999

INDEX

A

activism, 2–4, 24, 48, 117, 118–23, 125-27 (*See also* Navdanya; *satyagraha*); in agriculture, 6, 11, 90-91, 111, 118-19, 127; in courts, 3, 32, 49, 53-54, 68-69; global, 4, 18, 41-42, 117, 118-23, 125-27

A/F Protein, 51

Africa, 10, 21, 81, 83, 104, 125

Against the Grain (Bailey and Lappé), 100

Agracetus, 29, 81

agriculture, ecological, 14, 66, 75, 83, 101, 110-11 (*See also* farmers; seed; Navdanya); activism for, 90-91, 111, 118-19, 127; cows in, 57-59, 61-63, 67; in India, 3, 13, 48-51, 104-5; pest control, 97, 105, 106

agriculture, industrial, 1, 12, 105-6, 108, 117-18 (*See also* export agriculture; genetic engineering; herbicide; pest control); U.S. Department of Agriculture, 81, 82-84, 102, 110-11

Albrecht, Josef, 90-91

Al-Kabeer slaughterhouse, 68-69, 70, 77 (n 23)

Alvares, Claude, 3

amaranth, 21, 113-14

American Soybean Association, 25, 31

Andhra Pradesh, India, 10, 68-69, 101; activism in, 119, 122; shrimp farming in, 15, 48, 53

Andre & Company, 28-29

animal rights, 57, 69, 72, 74

antibiotics, 46, 66, 70, 102-3

apples, 29, 80

aquaculture. *See* fish

Archer Daniels Midland, 87-88

argemone, 24, 25, 26

Argentina, 28

ARISE, 119

Asgrow Seed, 29, 81, 91-92

Asia (*See also specific countries*): fish industry, 39, 43-44, 45, 46; grain, 21, 23, 81, 84; and World Trade Organization, 125

Astra/Zeneca, 81

Austria, 2, 120

Aventis, 81

130 STOLEN *Harvest*

B

Bacillus thuringiensis (Bt bacterium), 106-7, 109-10, 113, 122
Bailey, Britt, 100
Bangladesh, 44, 48
Basmati rice, 9, 85-86
Baucum, Scott, 92
beef. *See* cows
beets, 29, 105, 120
Benbrook, Charles, 113
Bengal, India, 22, 24, 49, 53, 104; famine in, 5-6
BGH (bovine growth hormone), 51, 60, 73, 100, 108
bheri aquaculture, 49
Bihar, India, 22, 24
biodiversity, 1-2, 18, 112-14, 123 (*See also* monoculture); food security and, 8, 79-80; genetic engineering and, 16, 95, 103-6; industrial agriculture and, 11, 13, 40, 62; traditional agriculture and, 21-22, 84-86, 99, 112; Convention on Biological Diversity, 90, 111
biosafety. *See* health concerns
biotechnology industry, 2, 81-82, 108, 111 (*See also* genetic engineering; multinational corporations); advertising by, 96, 97, 100-101, 120; genetic engineering and, 60, 82-84, 110, 113
Blue Revolution, 42-43, 51-52

Boeing, 125
Bollgard cotton, 92, 97, 99-100, 101, 119
Brazil, 18, 28, 37, 44
Britain, 5-6, 91; activism in, 2, 119; genetic engineering and, 51, 102, 105; mad cow disease in, 59-60, 63-65
British Society for Plant Breeders, 91
BSE (bovine spongiform encephalopathy, *aka* mad cow disease), 59-60, 63-65, 66, 70, 72, 73
BST (bovine somatrophin growth hormone), 51, 60, 73, 100, 108
Bt-crops (*Bacillus thuringiensis*), 106-7, 109-10, 113, 122
BUND, 120
Bunge, 28-29
Burrows, Beth, 122
Bush, George, 111
Business Ethics, 100
Business Line, 25

C

Calgene, 9, 26, 29, 81
Canada, 38, 92-93
canola, 29, 93, 104, 105
Cargill, 9, 16, 27, 28-29, 81, 87
carnivore/herbivore separation, 65, 72, 73
Carter, Jimmy, 27

turtles, 38-39, 40, 41-42

U

Unilever, 29, 81
Union of Concerned Scientists, 102
United Kingdom. *See* Britain
United Nations: Biosafety negotiations, 16; Convention on Biological Diversity, 90, 111; Food and Agriculture Organization, 38, 40, 67, 80; on shrimp farming, 15, 38, 40
United States, 45, 66, 71, 80, 118-19 (*See also* consumption); Alliance for Trade Expansion, 126; cows in, 58, 60, 76 (n 3); Department of Agriculture, 81, 82-84, 102, 110-11; Environmental Protection Agency, 106-7; environmentalism, 41-42, 70-71; export agriculture and, 9, 11, 15, 25-29, 39, 85; Food and Drug Administration, 109; genetic engineering in, 108, 110-12; multinational corporations and, 81, 88, 91-92, 111; Patent and Trademark Office, 86; patent law in, 89; Plant Variety Act, 92
Upanishads, 5, 12, 17
Upjohn, 60
Uttar Pradesh, India, 22, 24

V

vegetarianism, 21, 66-67
Venky's, 70
Verfaillie, Hendrik, 95, 97
violence, 11, 26, 60, 65, 74-75; against animals, 63, 72; suicide, 10, 101
Vishnu, 38

W

Wall Street Journal, 82, 108
water: conservation, 1, 105; salinization, 46-48, 50, 71
Weaver, James, 28
weeds, 16, 105-6 (*See also* herbicide)
Von Weiszacker, Christine, 122
wheat, 22, 26, 29, 81, 86-88, 113; biodiversity and, 12, 79, 80, 90-91, 103-4
White Revolution, 59-61, 60
Wills, Peter, 122
Winterboers, Becky and Dennis, 91-92
women: activism by, 3, 6, 15, 48; cows and, 58, 59, 60, 68; feminism, 72, 74-75; in traditional agriculture, 7, 16-17, 87, 104-5
World Bank, 10, 15, 42-44, 65
World Trade Organization (WTO): activism against, 125-26; Agreement on Agriculture, 2; dispute panel, 11,

Vandana Shiva is a world-renowned environmental thinker and activist. A leader in the International Forum on Globalization (IFG) along with Ralph Nader and Jeremy Rifkin and the Slow Food movement. Shiva won the Alternative Nobel Peace Prize (the Right Livelihood Award) in 1993.

Director of Navdanya and the Research Foundation for Science, Technology, and Natural Resource Policy, she is the author and editor of many books, including *Earth Democracy: Justice, Sustainability, and Peace* (South End Press, 2005), *Manifestos on the Future of Food & Seed* (South End Press, 2007), *Water Wars: Privatization, Pollution, and Profit* (South End Press, 2002) *Protect or Plunder? Understanding Intellectual Property Rights* (Zed, 2001), *Stolen Harvest: The Hijacking of the Global Food Supply* (South End Press, 2000), *Biopiracy: The Plunder of Nature and Knowledge* (South End Press, 1997), *Monocultures of the Mind* (Zed, 1993), *The Violence of the Green Revolution* (Zed, 1992), and *Staying Alive* (St. Martin's Press, 1989). Before becoming an activist, Vandana Shiva was one of India's leading physicists.

Related Titles

Earth Democracy: Justice, Sustainability, and Peace
Vandana Shiva

Water Wars: Privatization, Pollution, and Profit
Vandana Shiva

Stolen Harvest: The Hijacking of the Global Food Supply
Vandana Shiva

Biopiracy: The Plunder of Nature and Knowledge
Vandana Shiva

Manifestos on the Future of Food & Seed
edited by Vandana Shiva

Heat: How to Stop the Planet From Burning
George Monbiot

Toolbox for Sustainable City Living: A Do-It-Ourselves Guide
Scott Kellogg and Stacy Pettigrew

An Ordinary Person's Guide to Empire
Arundhati Roy

Power Politics
Arundhati Roy

War Talk
Arundhati Roy

Ecological Democracy
Roy Morrison

Earth for Sale: Reclaiming Ecology in the Age of Corporate Greenwash
Brian Tokar

Confronting Environmental Racism: Voices from the Grassroots
Robert D. Bullard, editor

¡Cochabamba! Water War in Bolivia
Oscar Olivera, in collaboration with Tom Lewis

What Lies Beneath: Katrina, Race, and the State of the Nation
edited by the South End Press collective

Recovering the Sacred: The Power of Naming and Claiming
Winona LaDuke

All Our Relations: Native Struggles for Land and Life
Winona LaDuke

Resource Rebels: Native Challenges to Mining and Oil Corporations
Al Gedicks

ABOUT SOUTH END PRESS

South End Press is an independent, collectively run book publisher with more than 250 titles in print. Since our founding in 1977, we have met the needs of readers who are exploring, or are already committed to, the politics of radical social change. We publish books that encourage critical thinking and constructive action on the key political, cultural, social, economic, and ecological issues shaping life in the United States and in the world. We provide a forum for a wide variety of democratic social movements and an alternative to the products of corporate publishing.

From its inception, South End has organized itself as a collective with decision-making arranged to share as equally as possible the rewards and stresses of running the press. Each collective member is responsible for editorial and administrative tasks, and earns the same base salary. South End also has made a practice of inverting the pervasive racial and gender hierarchies in traditional publishing houses; our collective has been majority women since the mid-1980s, and at least 50 percent people of color since the mid-1990s.

Our author list—which includes bell hooks, Andrea Smith, Arundhati Roy, Noam Chomsky, Vandana Shiva, Winona LaDuke, and Howard Zinn—reflects South End's commitment to publish on myriad issues from diverse perspectives.

read. write. revolt.
www.southendpress.org

Community Supported Agriculture (CSA) has helped to make independent, healthy farming sustainable. Now there is Community Supported Publishing (CSP)! By joining the South End Press CSP you ensure a steady crop of books guaranteed to change your world. As a member you receive all the new varieties and choice heirloom selections free each month and a 10% discount on everything else. (That is: every new book we publish while you are a member and selected backlist and bestsellers)

By becoming a CSP member you not only show solidarity with the collective that operates South End Press, but enable us to continue and expand the way we serve and support radical movement. CSP members are crucial to our ongoing operations. It is your regular support that allows us to keep your shelves stocked with important books (even if they aren't on the *NYT* bestseller list!).

Just as real change comes about from sustained grassroots organizing and people joining together, we know it is the modest support from a broad range of people that will allow South End Press to survive and grow.

It is only with the support of ordinary people on a regular basis that we can continue to struggle.

Won't you join the movement today?

To join please visit:
www.southendpress.org/2006/
items/80129